Hydrometeorology

Hydrometeorology

Edited by:

Pralhad Jaybhaye

Delve PUBLISHING

www.delvepublishing.com

Hydrometeorology

Pralhad Jaybhaye

Delve Publishing
224 Shoreacres Road
Burlington, ON L7L 2H2
Canada
www.delvepublishing.com
Email: orders@arclereducation.com

ISBN: 978-1-77469-137-3 (Hardcover)

ABOUT THE EDITOR

Dr. Pralhad Jaybhaye (1972) is presently serving as Assistant Professor, Department of Agricultural Meteorology, Vasantrao Naik Marathwada Krushi Vidyapeeth, Parhani, Maharashtra. He did his B. Sc.(Ag.) in 1995 and M. Sc. (Agril. Meteorology) 1998 from Mahatma Phule Krushi Vidyapeeth, Parbhani and Ph. D. (Agril. Meteorology &Physics) in 2020 from BCKV, Mohanpur, West Bengal. He started his career as a technical officer (Agrometeorology) in 2000 and as a Assistant professor in 2008. His field of specialization is Micrometeorology and Extension Agrometeorology. In addition to this, he has performed additional duties as a Nodal officer, Commercial Fruit Nursery (2005-2008) and Principal Nodal officer, GKMS & FASAL project (2008-20016). He was conducted National workshop, farmer training programmes, workshops and farmer's rallies. During his total 20 years' tenure of services, he has published one text book (Handbook of Agrometeorology) and other 6 books related to Agrometeorology; 25 research papers & 05 short communications in international and national reputed journals. He has participated and presented research papers more than 25 International, National, State Seminar, workshops and symposiums. He has written 01 practical manuals for under graduate course to the benefit of students; besides these activities TV (15) & Radio talks (200), guest Lectures in scientist training & farmer training programmes. Dr. Pralhad Jaybhaye has also written more than 400 popular articles in reputed regional language periodicals and daily news papers.

TABLE OF CONTENTS

LIST OF FIGURES

LIST OF TABLES

LIST OF ABBREVIATIONS

ATI	Area-Time Integral
CCN	Cloud Condensation Nuclei
CEH	Center for Ecology and Hydrology
CNES	Center National d'Études Spatiales
COPE	Convective Precipitation Experiment
COSMOS-UK	Cosmic-Ray Soil Moisture Observing System UK
DAD	Depth-Area-Duration
DD	Duration of Depth
DJF	December-January-February
DSD	Drop Size Distribution
ENSO	El Nio-Southern Oscillation
ESMR	Electrically Scanning Microwave Radiometer
ET	Evapotranspiration
EUMETSAT	European Organization for the Exploitation of Meteorological Satellites
FEMA	Federal Emergency Management Agency
FRP	Fiber Glass Reinforced Polyesters
FSG	Folland-Shaped Gauge
GCOM	Global Change Observation Mission
GEO	Group on Earth Observations
GEOSS	Global Earth Observation System of Systems
GEWEX	Global Energy and Water Cycle Experiment
GHG	Greenhouse Gas Emissions
GPCP	Global Precipitation Climatology Project
GPM	Global Precipitation Measurement
HSL	Health and Safety Laboratory
IGWCO	Integrated Global Water Cycle Observations
IMD	Meteorological Department of India

IPCC	Intergovernmental Panel on Climate Change
IR	Infrared
ISRO	Indian Space Research Organization
JJA	June-July-August
JTWC	Joint Typhoon Warning Center
LSPs	Land Surface Parameterization Schemes
METROMEX	Metropolitan Meteorological Experiment
MSS	Multispectral Scanner
NAO	North Atlantic Oscillation
NE	North East
NERC	Natural Environment Research Council
NHC	National Hurricane Center
NO	Nitric Oxide
NOAA	National Oceanic and Atmospheric Administration
NPP	National Polar-Orbiting Partnership
NRC	National Research Council
Pa	Pascal
PMP	Probable Maximum Precipitation
PMS	Probable Maximum Storm
PNA	Pacific-North American
POES	Polar-Orbiting Satellites
RH	Relative Humidity
SANDRP	South Asia Network on Dams, Rivers, and People
SAR	Synthetic Aperture Side-Looking Radar
SE	South East
SEBS	Surface Energy Balance system
SMMR	Scanning Multichannel Microwave Radiometer
SPS	Standard Project Storm
SVATs	Soil-Vegetation-Atmosphere Transfer Schemes
SWF	Surface Water Flooding
TDR	Time-Domain Reflectometry
TDT	Two Near-Surface Time-Domain Transmissometry
TIROS	Television Infrared Observation Satellites
TM	Thematic Mapper

TRMM	Tropical Rainfall Measuring Mission
UCM	Urban Canopy Model
UN	United Nations
WCRP	World Climate Research Program
WMO	World Meteorological Organization
WWTPs	Wastewater Treatment Plants

PREFACE

Our earth, also called the blue planet, contains water on its three-fourth surface while only one-fourth surface is dry (including deserts, mountains, and plain land). Water is essential for life on this planet. Nevertheless, extreme fluctuations in the water levels result in catastrophes in the shape of droughts and floods. The hydrological cycle is the circulation of water in the global atmosphere, and it is a key constituent of the planet's energy system. Human wellbeing is mainly influenced by the presence of water in the atmosphere. On the other hand, terrestrial surfaces significantly impact the atmosphere.

In recent times, the world has seen tremendous developments in the field of hydrologic and atmospheric sciences. However, climate predictions and meteorological predictions beyond a few numbers of days necessitate numerical simulation models that take account of realistic illustrations of surface hydrology along with energy exchanges on the surface. Therefore, it is indispensable that meteorologists and hydrologists work together. Thus, the field of hydrometeorology is significant, and is illustrated in this book. However, this field is so broad that this book is unable to encompass everything. Nevertheless, it is hoped that the book will stimulate the readers to explore the area of hydrometeorology further.

This book illustrates recent advances in hydrometeorological predictions, with an emphasis on modern forecasting techniques and meteorological observations. The topic of hydrometeorology includes a broad range of areas, such as rain gauge, weather radar, satellite, and other monitoring techniques, flow routing, hydraulic models, rainfall-runoff, and Mathematical Weather Predictions. Primary applications of hydrometeorology include drought forecasting, flood forecasting, reservoir management, climate change assessment, water quality, and water resources studies.

Topics of this book include an introduction to hydrometeorology, hydrological cycle, fundamentals of precipitation, evaporation, hurricanes, and tropical storms, meteorological measurements and instruments, satellite-based remote sensing, and hydrometeorology in urban environments. There are eight chapters in the book, and each chapter focuses on a particular area of hydrometeorology. Chapter 1 introduces the readers with the fundamentals of hydrology, meteorology, and hydrometeorology. Chapter 2 mainly focuses on the mechanisms of the hydrological cycle and associated processes. Chapter 3 offers a detailed discussion about the precipitation phenomenon and its varying characteristics. Chapter 4 sheds light on the impact of evaporation on hydrometeorology and hydrological cycles. Chapter 5 illustrates the science behind hurricanes and tropical storms and forms a linkage between hydrometeorology, hurricanes, and storms.

Measurement of water levels and intensities is essential for predicting the potential impacts of hydrometeorology on the environment. Chapter 6 focuses on the fundamentals of measuring gauges, their principles, and potential applications. Currently, satellites are being used to remotely sense the situation of a particular land. Chapter 7 elucidates the key concepts of satellite-based remote sensing for meteorological applications. Finally, Chapter 8 deals with the applications of hydrometeorology in urban environments.

The book offers valuable information for readers from different backgrounds, including civil engineering, meteorology, hydrology, and water resource technology. The book is equally beneficial for a general reader intending to equip themselves with the fundamentals of hydrometeorology.

—***Author***

Chapter 1

Introduction to Hydrometeorology

CONTENTS

1.1. INTRODUCTION

Water is one of the most important natural resources and essential for all kinds of living. It corresponds to the annual renewable portion of freshwater, which comprises surface water, underground water, and soil water. All fresh water on the planet comes from atmospheric precipitation (snowfall, rainfall, as well as some other forms). Rainfall, the main water resource that came as a response to atmospheric moisture condensation, is regulated by meteorology and is hence regarded as among the most significant meteorological factors (Baidya and Avissar, 2002). Likewise, the science of geology and hydrology governs the distribution and occurrence of rainwater on and below the surface of the earth. The hydrologic cycle or water cycle is associated with the flow and water exchange among oceans, the atmosphere, and the earth in both hydrology and meteorology. The meteorological aspects of evaporation and rainfall, along with hydrologic components of river and lake levels, groundwater, infiltration, and river flow, are the most noticeable parts of the hydrologic cycle that are commonly seen and documented for the sake of the sustainability of water. River stream data are particularly useful for measuring the availability of water within a river that can be kept in reservoirs for household, hydropower generation, irrigation, and some other uses (Phillips, 2012; Bruce and Clark, 2013).

The growth of a region's water resources, necessitates hydrological data, like flow of river, at a variety of locations across a river. In general, such statistics are not supplied for many of these rivers, particularly in underdeveloped nations, and observing them at a range of locations across a river is a costly and often challenging job. But at the other side, meteorological data like wind, rainfall, radiation, humidity, evaporation, vapor pressure, temperature, daylight hours, etc., are monitored continuously and archived over many years for a large network of stations across the globe (Wigley et al., 1984; Bruijnzeel et al., 2011). In the lack of true hydrological data, it is now widely accepted that meteorological data can be utilized to analyze and enhance a region's water resources utilizing hydrometeorology science. Hydrometeorology, then, is the field of meteorology that involves meteorology applicability to resolve hydrologic issues. In general, hydrometeorology is simply the analysis of the hydrologic cycle's land phases and atmospheric, with a focus on interconnections associated. It is a cross-disciplinary science that combines the study of the atmosphere, meteorology, with water-surface hydrology, the study of the occurrence, transport, distribution, and water storage also on the surface of the earth (Betts, 2004; Brown and Seo, 2010).

1.2. SCOPE OF HYDROMETEOROLOGY

Meteorology is the science dealing with the physics of atmosphere and the meteorology growing application to hydrological issues is directly tied to the growth of hydrometeorology as a science. The following examples were chosen to highlight various elements about the utilization of meteorological data in agricultural and hydrological studies. The field encompasses a wide range of topics, but we will focus on evaporation and rainfall in this section. Evaporation is frequently utilized to calculate losses from reservoirs, ponds, and, lakes as well as to estimate agricultural water needs. For hydrological modeling, evaporation estimates from basins (river) are employed (Fletcher, 1951; Varshneya and Balkrishana, 2003; Rakhecha and Singh, 2009). Surface runoff can be calculated by subtracting losses of evaporation from rainfall. The rate of evaporation is heavily influenced by meteorological factors such as daylight hours, temperature, cloud cover, humidity, vapor pressure and wind speed. These components of meteorology are extremely useful in indirectly calculating evaporation for hydrologic research where direct evaporation measurement is not feasible (Bleasdale, 1973; Wigley et al., 1984).

Data on rainfall is utilized in the preparation and building of water management projects. Building reservoirs and dams across rivers to store and collect river water runoff to meet people's needs has been a popular practice for ages. Data of long-period rainfall is utilized in the design and planning of these dams to estimate the flow of water in a stream. Data of stream flow is essential for determining the river basin's water production and determining the size of dam storage capacity. Dams are constructed and built with appropriate spillways to protect them from severe floods generated by heavy rainfall or thunder storm. This necessitates estimating the biggest floods based on outflow data which are frequently insufficient due to the scarcity of records and surface gauging stations (Erdil and Arcaklioglu, 2013; Kramareva et al., 2019). In such circumstances, a planned flood derived from a planned storm is used, based on the scope, placement, and type of the dam. A rainfall-runoff model is implemented to turn the design storm into a design flood based on a study of time rainfall information from stations in the region of river basin under examination. For calculating the design storm, many rainfall assessment techniques have been implemented, including storm transposition, depth-area-duration (DAD) and duration of depth (DD). The probable maximum storm (PMS) or probable maximum precipitation (PMP), the standard project storm (SPS) are two design storms that are mostly utilized to estimate design floods (Eskridge et al., 1997; Ishii et al., 2005).

The PMP's goal is to give an absolute safety criterion in cases where a dam's overtopping or failure would result in significant loss of property and life. Small dams are not designed to withstand the storm of PMP because a dam breach will pose little risk downstream. Alternatively, a frequency technique is utilized, in which severe rainfalls at a certain location are evaluated statistically to establish the frequency at which a particular intensity rainstorm for a set time can be forecast. The design flood is then estimated using hydraulic methods based on the storm rainfall designed (Xu and Singh, 1998; Apadula et al., 2012). The collected data from rain gauges are used to size channels of flow openings beneath bridges, stream control works, channels for railways and highways, urban storm sewers, airfield drainage, and other drainage projects.

All of them are created by taking into account the rainfall frequency and then using statistical ways to analyze the necessary rainfall intensity. Rainfall data and forecasts are crucial inputs for river flood alerts. The accuracy of rainfall forecasts is critical to the effective running of a multifunctional dam.

The amount of the input is calculated relying on rainfall predictions, as well as the reservoir is maintained to handle the oncoming water flood. Water use through dry seasons has to be properly managed in the case of storage reservoirs. For estimating water wave levels and determining freeboard needs, an evaluation of the likely wind across reservoirs is required (Ziomas et al., 1995; Smith, 2002; Zhi et al., 2021). Thus, necessity of hydrometeorological forecasting is the good quality of weather data with topographical real-time information and valid weather forecast, which can run hydrological models (Figure 1.1).

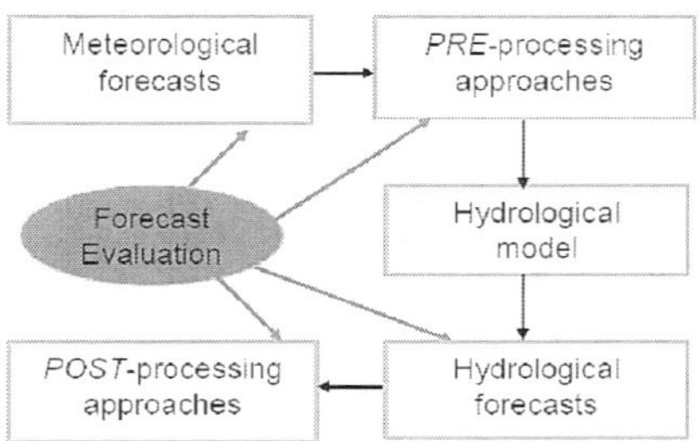

Figure 1.1. Diagrammatic illustration of hydrometeorological forecasting.

Source: *https://www.researchgate.net/figure/Schematic-of-a-hydro-meteorological-forecast-chain_fig1_258606377.*

1.3. METEOROLOGICAL VARIABLES

The atmosphere, vital to terrestrial life, envelops the earth to a thickness of only 1% of the earth's radius. The atmosphere is a fine layer of air that envelope the earth and thins as it rises. The condition of an atmosphere at a given location is represented by numerous meteorological factors such as pressure, humidity, wind, radiation, temperature, sunshine, rain, evaporation, clouds, fog, and so on at a certain time. The instantaneous atmospheric condition at a particular place and time is called weather and average condition of weather over a long period of time for the region or continent is called climate. Continuous monitoring of meteorological parameters is required to have a better knowledge of the atmosphere and its accompanying weather. Calculations are taken with standardized instruments and exposed to standardized conditions. The meteorological science is concerned with both weather origins and how it changes through time and space (i.e., variability, and change). Weather is not the same as climate (Thornton et al., 1997; Barry and Chorley, 2003; Xavier et al., 2016). Climate refers to the average weather of a location's which is calculated by combining meteorological data of many years (at least 30 years). The cause for this is that the trend of meteorological data variation through time replicates itself every 30 years or so (Flannigan and Harrington, 1988; Gocic and Trajkovic, 2013). The meteorological parameters related to hydrometeorology is required to make clear and which is described in subsections.

1.3.1. Temperature

Air temperature is the measure of degree of hotness of a body, whether solid, liquid or gas. The difference of temperature of two bodies indicates the direction of flow of heat and it is measured by different scale, which is elucidated in Table 1.1 (Jaybhyae et al., 2013). Temperature is a measurement of how hot or cold the air is, and it is measured using Fahrenheit or centigrade scale. The temperature of the earth is mostly determined by the quantity of radiation it receives. Aside from radiation, additional elements that affect temperature include prevailing winds, altitude, closeness to the sea, and so on. Temperature is measured in a Stevenson screen with a mercury thermometer for meteorological applications and pertains to the ambient temperature in shady situations at an altitude of around 1.5 meters. The temperature of the atmosphere is a very important parameter for calculating evaporation and snowmelt (Betts et al., 2005; Hamlet et al., 2005; Roy, 2011).

Lord Kelvin, while studying the gases' nature, the idea of absolute zero was developed, beyond which gases could not be cooled anymore. This temperature is around –273°C. This is where the absolute scale begins. The freezing temperature is 273°F, and the boiling temperature is 373°F. When dealing with extremely low or extremely high-temperature values, the Kelvin scale comes in handy (Marengo, 2009; Panthou et al., 2014; Yang et al., 2016).

Table 1.1. Different Temperature Scales Used for the Measurement of Temperature

SL. No.	Scale	Range	Melting Point of Ice	Boiling Point of Water
1.	Degree Kelvin	273°K–373°K	270°K	373°K
2.	Degree Celsius	0°C–100°C	0°C	100°C
3.	Degree Fahrenheit	32°F–212°F	32°F	212°F
4.	Degree Romar	0°R–80°R	0°R	80°R

1.3.2. Atmospheric Humidity

The forms of water in its different types are very important for all activities of life. Growth and sustenance of life heavily depends on it. The moisture content of atmosphere is about 0.035% of all fresh water. The water vapor decreases with increasing elevation or atmospheric moisture measured generally height in terms of relative humidity (RH). But moisture content of the air has been expressed in terms of the vapor pressure (Jaybhaye et al., 2013). The atmosphere acquires moisture by evaporation from oceans, lakes, rivers, and damp soil and from moisture transpired from plants. Taken together, these are often referred to as evapotranspiration (ET). Water vapor constitutes only a small fraction of the atmosphere, varying from near 0 to 4% of moist air by volume; it plays a major role in the heat budget and the weather phenomenon of the atmosphere. Evaporation through the world is predicted to be 1,000 (mm per year on average). Humidity relates to the quantity of vapor present in the atmosphere and it can be measured using a variety of physical parameters. The vapor pressure, RH, specific humidity, dew point, mixing ratio and absolute humidity, are all relevant humidity factors (Koutsoyiannis et al., 2007; Yamazaki et al., 2007; Varshneya and Pillai, 2018). The partial pressure of vapor present in the air is unaffected by the existence of many other gases. The vapor pressure, which is the most efficient estimation of the moisture level within atmosphere, is the pressure imposed by water vapor. It is typically represented by the letter "e" and it is estimated in millimeters or millibars of mercury, equivalent to atmospheric pressure. The quantity of vapor and the temperature of air that it can retain have an incredible relationship. The amount of vapor that must be present in the air is determined by the temperature of the air. More the vapor the air can carry, the greater the temperature (Lim et al., 2013; Brenot et al., 2014). At a given pressure and temperature, the air is considered to be saturated whenever it carries the greatest quantity of vapor possible. When vapor introduced is more or the air is chilled well down the saturation point, the vapor remain extra condenses into small dew droplets. The temperature at which dew forms and the pressure of vapor in the atmosphere becomes saturated is known as the dew point. The applied pressure by vapor in air saturated at a particular pressure and temperature is known as saturated vapor pressure, that is the maximum pressure of vapor for all practical uses. It is measured in millimeters or millibars of mercury and is represented by the letter "es." The saturation deficit is the difference of "es–e." The air's RH is stated as (Zhang et al., 2012; Snook et al., 2019):

$$RH = \frac{e}{e_s} \times 100 \qquad (1)$$

The percent ratio between actual vapor pressure (e) and the maximum feasible vapor pressure (es) over a specific temperature is thus described as RH. Whenever the RH becomes 100%, the air becomes saturated with moisture.

The following is the relationship between saturation vapor pressure and air temperature:

$$e_s = 6.107 \times 10^{\frac{7.5T}{(237+T)}} \qquad (2)$$

where; "es" denotes the saturation vapor pressure(millibars) and T denotes the temperature in degrees Celsius. For various temperatures, Eqn. (2) could be used to calculate the saturation vapor pressure. Table 1.2 shows the saturation vapor pressure values for different temperature levels.

The utilization of dry and wet bulb thermometers is perhaps the most practical way for measuring air RH. A thermometer bulb covered in a smaller strip of muslin that is maintained wet by a wick placed in a little water bottle should register a lesser air temperature than just an equivalent bare thermometer. A psychrometer is made up of two similar thermometers set side by side. When the overall RH of air reduces, the reading of the wet-bulb thermometer decreases. The lower temperature reading out from the wet-bulb thermometer is attributed to the water evaporation from the muslin (in wet form) into the surrounding atmosphere whenever the air is unsaturated. The air is cooled as the latent heat of vaporization is extracted. The air continues to cool and the moisture content of the air increases due to evaporation from muslin till the wet bulb of air surrounding is saturated. Following that, a thermometer of a wet-bulb consistently records a lower value than the dry bulb. The difference seen among wet bulb and dry bulb thermometer readings regarding the atmospheric temperature provides a measurement of the real RH as well as the vapor pressure. For measuring vapor pressure and RH equations, diagrams, and tables have been produced.

Table 1.2. Vapor Pressure at Saturation (mb)

Temperature (°C)	**Saturation Vapor Pressure (mb)***									
	0	**1**	**2**	**3**	**4**	**5**	**6**	**7**	**8**	**9**
00	6.11	6.57	7.06	7.58	8	8.72	9.	10.02	10.73	11.49
10	12.29	13.14	14.03	14.99	16.00	17.07	18.20	19.39	20.66	22.00
20	23.41	24.90	26.45	28.13	29.88	31.73	33.67	35.71	37.87	40.13
30	42.51	45.01	47.65	50.40	53.31	56.35	59.50	62.89	66.40	70.08
40	73.94	77.97	82.20	86.62	91.24	96.08	101.14	106.42	111.94	117.70

*To acquire the value of proper saturated pressure of vapor from the table regarding a particular temperature, check the temperature in the first column, then value would be added to the very first row. Go to the 4th row, that represent a number of 20, and the eighth column, that refers to a number of 6, and the similar element of 33.67 mb, which is the saturated vapor pressure for 26°C.

The psychrometric equation for calculating the air vapor pressure is as follows:

$$e = e_s - 0.000660\, P\, (T - T_w)\,(1 + 0.00115\, T_w) \qquad (3)$$

where; e is the vapor pressure in the air expressed in millibars; P is the air pressure in millibars; e_s is the saturated vapor available in millibar at the wet-bulb thermometer's temperature Tw in °C; T_w is the wet-bulb thermometer temperature in °C; *T* is the dry bulb thermometer temperature in °C

For a 1,000 mb pressure, Table 1.3 shows the vapor pressure (e) values that correspond to specific values of the dry bulb and wet bulb temperatures.

Humidity can be show in a variety of ways, including a wet-bulb thermometer, RH, and vapor pressure, as discussed below.

Vapor density is the amount of vapor present in a unit volume of moist air. This is also known as absolute humidity. It is calculated in grams per square meter (gm/m^3). The vapor mass contained in a gram of wet air is indicated by the letter "q." It is rarely higher than 0.04.

The mixing ratio is indicated by the letter "r." and is defined as the water vapor mass contains in gram of dry air. The mixing ratio and specific humidity are commonly stated in grams per kilogram.

Table 1.3. Values of e (mb)

T_w (°C)	$T - T_w$ (°C)									
	0	**01**	**02**	**03**	**04**	**05**	**06**	**07**	**08**	**09**
0	6.1	5.4	4.7	4.1	3.1	2.7	2.0	1.4	0.7	0.08
2	7.0	6.2	5.7	5.0	4.3	3.7	3.0	2.3	1.6	1.0
4	8.1	7.4	6.7	6.1	5.4	4.7	4.0	3.4	2.7	2.0
6	9.3	8.6	8.0	7.3	6.5	5.9	5.3	4.6	3.9	3.2
8	10.7	10.0	9.3	8.7	8.0	7.3	6.6	6.0	5.3	4.6
10	12.2	11.6	10.9	10.2	9.5	8.8	8.2	7.5	6.8	6.1
12	14.0	13.3	12.6	12.0	11.2	10.6	9.9	9.2	8.6	7.9
14	16.0	15.3	14.6	13.9	13.2	12.5	11.9	11.2	10.5	9.8
16	18.2	17.5	16.8	16.1	15.4	14.6	14.5	13.4	12.7	12.0
18	20.6	19.4	19.2	18.6	17.9	17.2	16.5	15.8	15.1	14.5
20	23.4	22.7	22.0	21.3	20.6	19.9	19.2	18.6	17.9	17.2
22	26.4	25.8	25.0	24.4	23.7	23.0	22.3	21.6	20.9	20.2
24	29.8	29.1	28.4	27.8	27.1	26.4	25.7	25.0	24.3	23.6
26	33.6	32.9	32.2	31.5	30.8	30.2	29.5	28.8	28.1	27.4
28	37.8	37.1	36.4	39.7	35.0	34.3	33.6	33.0	32.3	31.6

When P indicates the total air pressure and e denotes the partial pressure of vapor contained in it, the dry air partial pressure equals (P – e). The mixing ratio(r) and specific humidity (q) are calculated using the following formula:

$$r = \frac{\varepsilon e}{P - e} = \frac{0622e}{P - e} \tag{4}$$

where; 0.622 is the vapor molecular weight divided by the molecular weight of the dry air. Tables 1.2 and 1.3, as well as Eqns. (2) and (3), can be used to calculate humidity parameters (3).

$$q = \varepsilon \frac{e}{P} = \frac{0622e}{P} \tag{5}$$

1.3.3. Atmospheric Pressure

As previously stated, a gaseous envelope encircled the world from all sides as the termed as an atmosphere, that extends several kilometers just above the earth's surface. Since air possesses weight, an imagined tube having unit area(cross-sectional) filled with air that expands from the ground to its topmost limit would exert a certain pressure on objects present on the earth surface. The force exerted on a unit area by an air in a Column is known as atmospheric pressure, and it unite in the CGS system is dynes/cm^2. If the column of height *h* at mean sea level of air with an average density of, then perhaps the 'gh' is the atmospheric pressure (p), while 'g' is the gravitational acceleration (Cleverly et al., 2013; Neiman et al., 2013).

A barometer based on mercury discovered by Torricelli (1643) is used to measure pressure. The mercury column height in a barometer is used to expressed pressure while balancing atmospheric pressure. The mercury column height at sea level is almost 76 cm. Because mercury has a density of 13.59 g/cm^3 at 15°C and an acceleration of 980 cm/s^2, the sea level average pressure is 1,013,200 dynes/cm^2. The dynes/cm^2 unit is extremely tiny. The millibar, which is 1,000 times greater, was introduced by Bjerknes. In meteorology, it is a pressure unit. The average pressure of sea level is 1,013 mb. In the SI system of measurement, 1 N/m^2 is the unit of pressure. The Pascal (Pa) is a unit of measurement that equals 10 dynes/cm^2. As a result, 1 mb = 100 Pa (also known as 1 hectopascal or 1 hPa). In the SI system, the atmospheric pressure of 1,013 mb equals 1,013 hPa (Vivoni et al., 2007; Jaybhaye et al., 2013; Hao et al., 2018).

1.3.4. Air Density

The density of air represented as-is on the scale of 1/1,000 that of water close to the ground and is practically never determined. It is instead calculated using the equation of gas ($\rho = P/RT$). It is expressed in kg/m^3 or gm/cm^3 units. The density of air is 0.0012 gm/cm^3 at a 1,000 mb surface pressure and at 290°K temperature (Snook et al., 2019).

1.3.5. Wind

Air in horizontal motion is known as wind. This motion tends to equalize lateral differences in temperature, humidity, and pressure in the atmosphere. This equalization is never attained since new differences are created. Vertically moving air columns are called currents. Average horizontal wind speeds are much greater than average vertical movements. The science of measuring and recording the direction and speed of wind is called 'Anemometery.' Instruments for surface wind speed measurement are known as 'anemometers,' while, those that make a record are 'anemographs.' Devices which measure and record wind direction alone are known as 'wind vanes' (Jaybhaye et al., 2013).

1.3.5.1. Horizontal Winds

A wind that is the longitudinal movement of air on the earth's that is defined by its speed and direction. The wind direction is determined by the way the wind blows. The cardinal directions, like N, NNW, NW, and so on, are used to describe the wind directions (Hunter et al., 2015). Wind direction is the direction from which the wind is approaching the observer. It is expressed in degrees, measured clockwise from geographical north. As it is average value, it is normally expressed in codes from 00 to 36 where 00 represents calm, 09 easterly, 18 southerly, 27 westerly 36 northerly and so on. The direction of wind is also expressed in 8 or 16 points of compass. Expressing north, east, south, and west by N, E, S, and W, respectively, these directions represented by 8 and 16 points of compass are:

- **8 Points of Compass:** N, NE, E, SE, S, SW, W, NW, N representing direction from 3,400 to 200
- **16 Points of Compass:** N, NNE, NE, ENE, E, ESE, SE, SSE, S, SSW, SW, WSW, W, WNW, NW, NNW: N representing direction from 3,500 to 100. It is measured with the help of a wind vane instrument and the north wind (N) is a wind that blows from the north. Speed of wind is expressed in kilometers per hour, meters per second (meter per second), miles per, or knots (nautical miles per hour). The wind vane is used to determine the direction of the

wind, while the anemometer calculate wind speed. For winds, the conversion factor is: knot = 1.853 km/hr = 1.57 miles/hr = 0.51 m/s (Jaybhaye et al., 2013).

1.3.5.2. Vertical Motions of Air

The majority of the cooling and subsequent condensation of air, which leads to cloud formation and precipitation, is caused by the upward perpendicular motion of air on both a large and small scale. The velocity does not remain constant vertically and has a logarithmic distribution or power law up to a certain altitude (Hess et al., 1998).

1.3.6. Clouds

In hydrometeorology, precipitation, and clouds play a significant role. At higher elevations, supersaturated anabatic wind causes clouds to form. Just like air rises steadily to the earth's atmosphere, it expands, causing cooling. Evaporation of surface waters such as rivers, oceans, lakes, and so on creates water vapor, as previously stated. Because water vapor is light, it rises to the top of the atmosphere, where it swells adiabatically and cools just under the saturation point. Excess moisture condenses as minute droplets of liquid on suspended particulate matter, forming clouds (Plummer and Maynard, 2014).

Cloud is a visible of minute particles of water or ice or of both in the free air and Clouds are divided into three groups based on their appearance when seen from the surface (Figure 1.2): (i) layer clouds or stratus, (ii) heap-shaped clouds or cumulus, and (iii) feathery streaks of clouds or cirrus. Clouds are categorized as a medium, high, or low depending on how height above the surface they happen. The 10 major categories are listed below and shown in Table 1.3 per the international categorization (Brunsell and Marcks, 2005; Varshneya and Pillai, 2018). It may classify previously by meteorologist into 10 main cloud types (viz., cirrous; cirro cumulus; cirrostratus; altocumulus; Alto stratus; nimbostratus, stratocumulus, stratus, cumulus, and cumulonimbus) which are grouped into 3 families (i.e., high clouds; medium clouds and low clouds) based on altitude. Now a day clouds classified into 10 main cloud types as mentioned above, which are grouped into 4 families (viz., high clouds; medium clouds; low clouds and vertical clouds) and in addition to this fog; special clouds; earth quake clouds; bursting clouds, industrial clouds and launch vehicle (Agniban) clouds are considered (Jaybhaye et al., 2013). Cirrus clouds have a fibrous structure and appear as if they are made of glass wool drifting across a blue sky. A thin sheet of elevated clouds known as Cirrostratus gives the sky a milky appearance. Cirrocumulus is made up of a particular structure of white globules (Allison, 1997). Altostratus is a closely packed grey cloud with a sliver of the sun visible through it. Altocumulus is made up of globular masses that are often assembled in parallel bands with straightforward voids in between. Stratus is a low cloud that appears like the fog that has been lifted over the surface of the earth. Massive lumpy masses with disparities make up stratocumulus. Rain falls from Nimbostratus, a dense dark cloud mass.

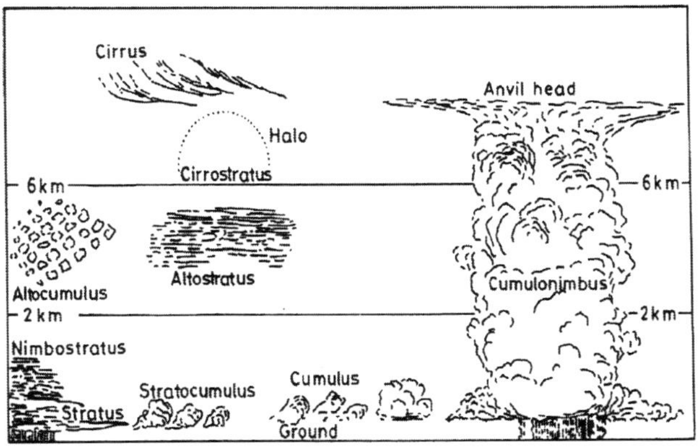

Figure 1.2. Clouds categories.

Source: *https://www.sciencedirect.com/science/article/pii/B978008011714050 0065.*

1.3.6.1. Measurement of Cloud

The amount of clouds is usually estimated visually. If in patches, mentally these are taken to be one patch and an estimate made as to how much of the sky the total cloud patch would cover. *The amount of cloud is reported in Oktas or eighth of the sky covered by the clouds.* When observing, the amount of each type of cloud present in the sky is recorded separately, as also the total coverage of the sky by all the cloud put together, total cloud amount of 8 oktas would mean an overcast sky (refer Jaybhaye et al., 2013, for detail information of measurement of cloud, and meaning in terms of meteorology). The height of the base of the clouds is estimated or measured by optical methods. Nowadays modern technology is also helpful to measure the cloud and to identify the types of cloud, in which radiosonde, radar, Doppler radar and satellite is used. Even though by remote sensing technology photographs of cloud globally regionally is receiving. From these photographs meteorologist and also common people can able to classify the types of cloud. By this technology, meteorologist estimate the cloud cover, types of cloud and direction of cloud to be useful for rainfall forecasting. Also, it is useful in artificial cloud seeding, i.e., artificial rainfall experiment (Jaybhaye et al., 2013).

1.3.7. Precipitation

Precipitation, condensation, and evaporation are the three central elements of the water cycle within the atmosphere. Precipitation is the word applied to represent the water that drops on the surface of the earth from the atmosphere in either solid or liquid form. Rain in a state of liquid as well as sleet, hail, and snow in form of solid are all included. Water vapor evaporates from the surfaces of water such as ponds, seas, rivers, and lakes, as well as from vegetation and land. When water vapor reaches significant altitudes well above the earth's surface, it forms clouds by condensation. Condensation is defined as the collection of water vapor molecules into extremely tiny droplets (Zhang et al., 2012; Lim et al., 2013).

These little water droplets develop in size and eventually fall as snow or rain on the ground. A raindrop has a radius of 1,000 microns on average. Precipitation must be measured to determine the quantity of water collected from the atmosphere in a specific region over a certain period. All types of precipitation are quantified in units of the vertical water depth which would collect on level ground if the complete precipitation stayed in place. Rain gauges are used to measure the two major components of precipitation (snow and rain).

The receiver on a rain gauge (standard non-recording) is five inches in diameter The most significant rainfall data for water resource development and hydrologic objectives are the rainfall mean depth for distinct places across different periods, like seasons, months, or the entire year. The average yearly precipitation in the world is roughly 80 centimeters. However, towards the equator between about 40° and 50° latitudes, precipitation is at its maximum. Over mountainous areas, there is unusually strong rainfall. The average annual precipitation in Cherrapunji [25.3°N, 91.8°E] of India is 1,087 centimeters (Baidya Roy, and Avissar, 2002; Li et al., 2021).

1.3.8. Evaporation

Evaporation is the transformation of a material from a liquid state to a vapor state. The heat of the sun causes the vaporization of surface water from lakes oceans, rivers, and other bodies of water, resulting in water vapor content in the atmosphere.

The Pan-measurement method, also known as the weather pan, is used to determine the rate of evaporation. This process involves using a pan with standard dimensions and filling it with water to a specific level. The quantity of water that has evaporated from this pan has been calculated, and the rate of evaporation per unit area per unit time has also been measured (Pellarin et al., 2002; Jaybhaye et al., 2013).

1.3.9. Visibility

The visibility of an object is described as the maximum parallel distance over which it can be viewed. It is calculated by choosing several visibility marks, like tall buildings, towers, and hills.

The visibility would be infinite if there were no particles in suspension. However, there are always a few particles in the suspended form in the atmosphere, reducing visibility. While, dust, smoke, and water particles are small enough to obscure visibility. Haze is created by dust particulates or smoke, while mist or fog is created by water particles. Heavy drizzle, heavy rains or snow, sandstorms, and Dust storms can turn down visibility to just a few 100 meters (Karyampudi and Carlson, 1988).

1.4. SEASONS AND EARTH ORBITAL MOTIONS

The radiation input varies in space and time due to the earth's inclination of the axis, revolution around the sun, and rotation around its axis. As a result, the earth has various seasons and climatic zones (Smithsonian Institutions, 1939; Sulis et al., 2019). The earth spins on its axis in an elliptical orbit at 113,000 kilometers per hour, taking per year to complete the 966-million-kilometer journey.

The earth's (E) orbit around the sun (S) is illustrated in Figure 1.3. The earth is said to be at perihelion when it is at P in its orbit around the sun. On or about January 1st, this happens. The earth is said to be at aphelion because it is at (A) and is the furthest away from the sun. This happens on or about July 2nd.

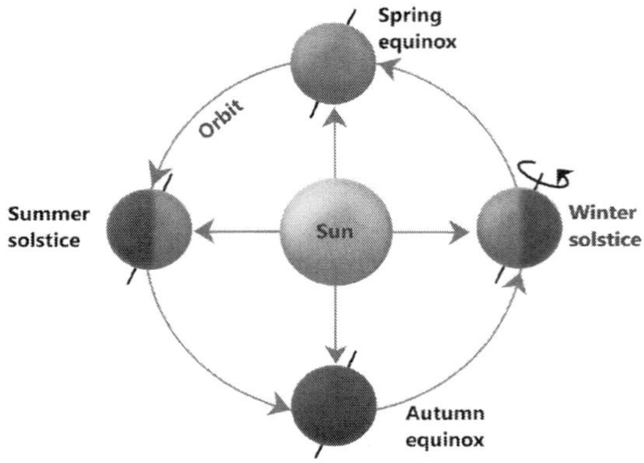

Figure 1.3. The earth's orbit around the sun.

Source: https://www.researchgate.net/figure/The-schematic-Earth-orbit-about-the-Sun-48-shown-not-to-the-real-scale-of-the-Sun-and_fig5_333988067.

The mean length among the earth and sun, or the earth's orbit semi-major axis, is the unit of astronomical distance. It has a value of 149.6×10^6 km, like, $OA = OP = 149.6 \times 10^6$ km. The earth's eccentricity (ec) path is 0.0167 at present. As (S) is the elliptical focus. We got x.

$OS = OP \times ec$ $149.6 \times 10^6 \times 0.01675 = 2.510^6$ km

$SP = 149.6 \times 10^6 - 2.510^6 = 147.1 \times 10^6$ km

$SA = 149.6 \times 10^6 + 2.510^6 = 152.1 \times 10^6$ km

The proclivity of the planet's equator to the plane of its trajectory around the sun is responsible for the winter and summer seasons. The planet's equator is oriented at a 23.5° angle to its orbital plane. This means that as the planet (earth) orbits the sun, the proportion of sunlight falling on it is unequal and varies (Figure 1.4).

The earth spins once every 24 hours around its elliptical orbit. The earth spins from east to west, and its axis sustains a tilt of 23.5° while it spins and revolves. The sun rotates all-around ecliptic and passes the equator 2 times per year when regarded from Earth. On or around March 21st, the vernal equinox occurs at the equator's point. Whenever the sun passes the equator from north to south. The autumnal equinox, which falls on or about September 22nd, is the other north-to-south crossing. Now or then, the sun rises at midday on the equator. While both day and night have the same period-12 hours-all over the globe (Rao, 1975; Khavrus and Shelevytsky, 2010).

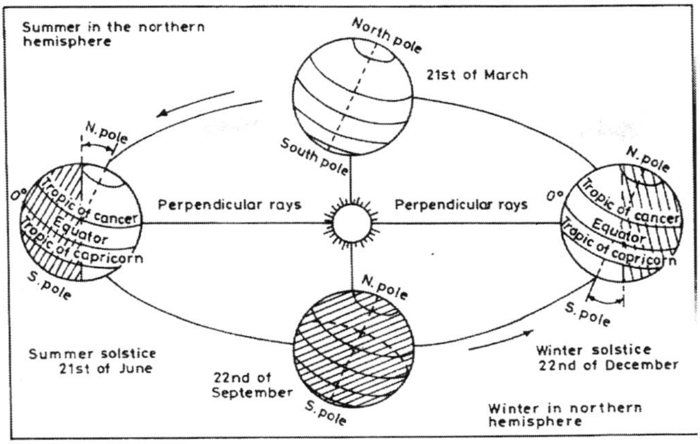

Figure 1.4. Winter and summer seasons are determined by the earth's equator inclination to its plane about the sun.

The sun starts to move northward later passing the vernal equinox on March 21st, attaining the threshold steady decline of 23½°. This happens on June 21st. That day-on the latitude circle of 23.5°N-the sun rises above the horizon at regional midday. While the north pole experiences the year's shortest night and longest day. On June 21st, the summer solstice occurs in the north pole. But on the same day, the winter solstice occurs in the south pole. Solar radiation reaches its highest point in the north pole, while it reaches its lowest point in the south pole. At 66½° north latitude, the day is 24 hours long on June 21st. To the north of the 66½°N parallel of latitude, the timeframe of light continues to increase, and light usually takes 6 months in the northern hemisphere. In the south pole, it is the reverse at this time. The sun does not shine in the south pole. Light reaches only around a portion of the south pole. On the 21st of June, night lasts 24 hours at 66½°S parallel of latitude. From the south of the 66½° parallel of latitude, the timeframe of darkness increases. whereas the southern hemisphere remains dark for 6 months (Petterssen, 1969; Yu et al., 2015).

The sun's superficial south-ward rotation begins with the summer solstice. Which crosses the equator at the autumnal equinox, and continues southward. On December 22nd, the sun reaches its southernmost declination of –23.5°. The northern hemisphere's winter solstice falls on December 22nd. Whereas, the southern hemisphere's summer solstice falls on the same day. Thus, solar radiation reaches its peak in the southern pole, though it reaches its lowest in the northern pole (L'vovich, 1979; De Paor et al., 2017).

1.5. WATER AND LAND GLOBAL DISTRIBUTION

The Earth surface is uneven. The earth surface is made up primarily of ice, ocean, and land, with the former covering only a little portion of the planet's surface (3%). The physical properties of these three components are very different. As a result, the distribution of ocean and land across the globe has a significant impact on climatic and weather management systems. The land covers 29.2% of the earth's surface, while water covers 70.8%. The statistics are 60% (water) and 39% (land) in the northern hemisphere, and 80% (water) and 19% (land) in the southern hemisphere. Furthermore, the land area fluctuates from one latitude belt to the next, as well as from one hemisphere to the next (CGWB, 1991; Sterling et al., 2013). The percentage land distribution fractions for varying 5° latitudinal belts are shown in Table 1.4.

Table 1.4. Clouds Classification, Their Symbol, Height (asl) and Characteristics

Cloud Name	Symbol	Occurrence Height	Remarks
Altostratus	As	2–6 km	They consist of super-cooled water droplets giving a precipitation as a light drizzle or snow.
Altocumulus	Ac	–	They are high globular groups are sometimes referred to as 'sheep' clouds or woolpack' clouds. They consist of ice crystals, super-cooled water droplets or a mixture of both.
Stratocumulus	Sc	< 2 km	These are large globular masses or rolls of soft gray clouds.
Stratus	St	–	These are without any particular form or structure. They are made up of water droplets giving a dull overcast sky.
Nimbostratus	Ns	–	They are often difficult to see since they continuously produce precipitation.

Cumulus	Cu	Cloud with great vertical development	It is generally found in the daytime over land areas. It dissipates at night. Most of the cumulus clouds are fair-weather type.
Cumulonimbus	Cb	–	This type of cloud is associated with heavy rainfall, thunder, lightning, and hail. It has an anvil head and flat base.
Cirrus	Ci	6–15 km	The sun or moon shining through these clouds produces a halo. They show brilliant colors at sunset and at sunrise.
Cirrostratus	Cs	–	They commonly produce a halo around the sun or moon.
Cirrocumulus	Cc	–	It is usually without shadows.

The most important aspects of this apportionment are (Tateishi and Ahn, 1996; Nakaegawa, 2012):

i. The percentage of land in the southern pole is 20–25% up to 30°S, then drops to zero by 50°S, including almost no property in 50°–65°S, and then rises sharply to 100% by 80–90°S.

ii. In the northern pole, the percentage rises from about 20% closer to the equator to around 70% in the belt between 65° and 70°N latitude, before falling sharply to zero near the north pole.

The latitudinal belts from the equator to 30° latitude (tropical), 30° to 60° latitude (mid-latitude), and 60° to 90° latitude (high latitude zones) are in the two hemispheres, respectively, for meteorological reasons (Table 1.5) (Ackermann et al., 1955; Cao et al., 2014).

Table 1.5. Land Area Distribution by Latitude [S = South and N = North]

Latitude Belt	**Percentage of Land (%)**		**Latitude Belt**	**Percentage of Land (%)**	
	N	**S**		**N**	**S**
0–05°	21	22	45–50°	56	03
05–10°	24	25	50–55°	59	02
10–15°	23	24	55–60°	55	00

15–20°	29	20	60–65°	69	00
20–25°	35	23	65–70°	71	20
25–30°	40	24	70–75°	35	61
30–35°	42	16	75–80°	23	69
35–40°	43	07	80–85°	23	69
40–45°	49	04	85–90°	00	100
			Total Land	**39**	**19**

1.6. BUDGET OF GLOBAL WATER

Table 1.6 illustrates an approximation of the global water balance. According to the data, the total volume of water on the planet is approximated to be 1.36 × 10^9 km^3. Approximately 94% of this is in the form of oceanic seawater (Vörösmarty et al., 2000; Milly and Shmakin, 2002; Pfister et al., 2011). The majority of the remaining water (about 2%) is locked up in glaciers and polar ice caps, with a small amount remaining as subsurface water (less than 0.01%). Groundwater makes up approximately 4% of the total. Freshwater exists in rivers and lakes in such small amounts (0.01%) that it can be used for domestic, industrial, and agricultural purposes. Approximately 0.001% of the world's water is in the form of clouds in the atmosphere. Atmospheric procedures play a critical role in replenishing the world's fresh water supply by evaporating from the oceans, condensing the vapor into clouds, and transporting moisture via air currents to distant locations, which then precipitates as snow or rain (Byers, 1959; Still et al., 2003; Beer et al., 2010).

Table 1.6. Balance of Global Water (L'vovich, 1970)

Location	**Volume of Water × 10^6 km^3**	**Global Water Percentage**	**Fresh Water Percentage**
Oceans	1,278	94	–
Polar ice and glaciers	27	2	33.33
Groundwater	54	4	66.66
soil water	0.08	0.0066	–
Lakes and rivers	0.13	0.011	0.166
Atmosphere	0.013	0.0011	–
Total global water	1360.0	100.0	–
Freshwater	81.73	6	100.0

1.7. INDIAN WATER RESOURCES

It is fascinating to learn about the water balance in a specific region of the world, like India. India is the world's seventh-largest country and the second largest in Asia, with a total land area of 3.29 million square kilometers. The country is situated within the Asia and comes under the monsoon belt, which is known for its heavy rainfall. The country receives about 117 cm of rain on average each year (Dhar and Rakhecha, 1979; Nernec, 1983). This is nearly 1.5 times greater than the global average of 80 centimeters per year. The total volume of water in India, calculated by multiplying rainfall by the country's land area, is around 4,000 km^3. About 1,400 km^3 of water is lost to evaporation, while 720 km^3 is absorbed by the soil. The average surface overflow in the country's river systems is projected to be 1,880 km^3 after accounting for infiltration and evaporation (Rao, 1975; Sahoo et al., 2011). The Central Ground Water Board, New Delhi (1991), calculated a 432 km^3 recharged India's water-table aquifer from rainfall. The India's surface water potency is approximately 1,880 km^3, which is distributed across the country's various river basins. These river basins provide water to over 1,000 million Indians. The water accessibility per person is 1,880 cubic meters per year, and it is about $1/12^{th}$ of the global average of 21,800 cubic meters per year for each person. India's population is expected to be between 1.5 and 1.8 billion people in 2050, according to estimates. Water accessibility per capita will be about 1,200 m^3/yr, putting India on the verge of becoming a water-stressed nation (Chen et al., 1998; Trenberthet al., 2007). Tables 1.7 and 1.8 illustrate the water resources of India overall, as well as the water resources of different Indian river systems. A map of India with river basins can be seen in Figure 1.5. There are 14 major river basins in India. The important Himalayan rivers in northern India are the Brahmaputra, the Ganga, and the Indus. Because these river systems are both snow- and rain-fed, they have a year-round flow. The Ganga River and its side streams stretched out as a fan across India's plains, developing the country's largest river basin, covering roughly a quarter of the country's total land area. The Cauvery, Mahanadi, Sabarmati, Subarnarekha, Krishna, Tapi, Godavari, Pennar, Mahi, and Narmada are the major rivers in southern and central India. These rivers are exclusively dependent on rainfall (Liu et al., 2010; Wong et al., 2011).

The irregular share of these estuary basins' water resources is one of their most striking features. The Indus, Brahmaputra, and Ganga River basins account for 65% of total drainage in the nation, while the Narmada, Krishna, Tapi, Godavari, and Mahanadi River basins account for only 17% (Table 1.8).

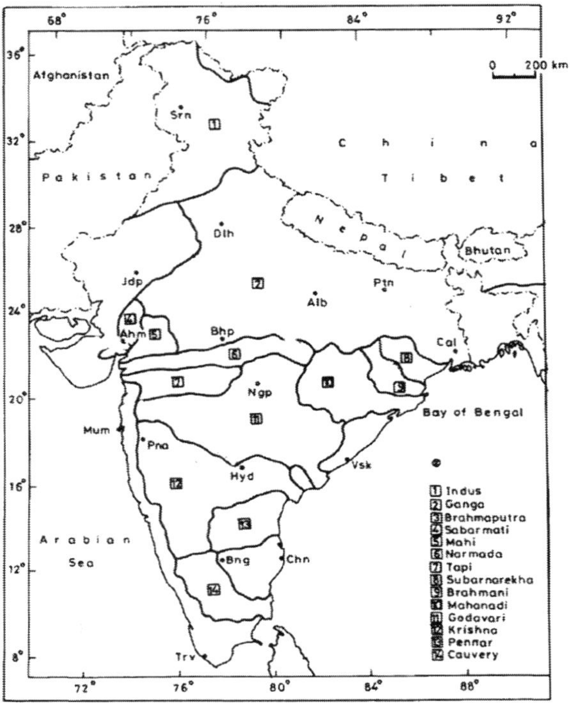

Figure 1.5. India's major river basins.

Source: https://www.sciencedirect.com/book/9780080117140/introduction-to-hydrometeorology.

Table 1.7. India Water Resources

Item	The Volume of Water (km^3)
1. Annual rainfall over India	4000.0
2. Evaporation loss	1400.0
3. Runoff from rainfall	1880.0
4. Seepage into the subsoil (1–2–3)	720.0
5. Soil moisture	288.0
6. Groundwater recharge (4–5)	432.0

Table 1.8. The Average Yearly Surface Runoff of India's Major River Basins

SL. No.	River Basin	Runoff in km^3	Percentage of Total (%)
1.	Indus	80.0	4.20
2.	Ganga	550.0	29.30
3.	Brahmaputra	591.0	31.40
4.	Sabarmati	4.0	0.20
5.	Mahi	11.0	0.60
6.	Narmada	40.0	2.10
7.	Tapi	20.0	1.10
8.	Subarnarekha	12.0	0.60
9.	Brahmani	29.0	1.50
10.	Mahanadi	67.0	3.60
11.	Godavari	116.0	6.20
12.	Krishna	58.0	3.10
13.	Pennar	7.0	0.40
14.	Cauvery	19.0	1.00
15.	West flowing rivers of Kutch-Saurashtra including Luni	13.0	0.70
16.	West flowing rivers south of Tapi	218.0	11.60
17.	East flowing rivers between Pennar and Cauvery	23.0	1.20
18.	East flowing rivers between Mahanadi and Godavari	22.0	1.20
	Total Surface Runoff	**1880.0**	

Despite the fact the average annual water flow in multiple rivers is 1,880 km^3, climate, soil conditions, and topography confine the amount of water that can be used to only around 720 km^3. In India, there are approximately 3,000 dams with a total storage capacity of 132 km^3 (Stohl and James, 2004; Takle et al., 2007).

REFERENCES

1. Ackermann, W. C., Colman, E. A., & Ogrosky, H. O., (1955). From ocean to sky to land to ocean. *In: U.S. Department of Agriculture Yearbook* (Vol. 1, pp. 1–23). U.S. Department of Agriculture, Washington, D.C.
2. Allison, M., (1997). Accurate analytic representations of solar time and seasons on Mars with applications to the Pathfinder/Surveyor missions. *Geophysical Research Letters, 24*(16), 1967–1970.
3. Apadula, F., Bassini, A., Elli, A., & Scapin, S., (2012). Relationships between meteorological variables and monthly electricity demand. *Applied Energy, 98*, 346–356.
4. Baidya Roy, S., & Avissar, R., (2002). Impact of land use/land cover change on regional hydrometeorology in Amazonia. *Journal of Geophysical Research: Atmospheres, 107*(D20), LBA-4.
5. Barry, R. G., & chorley, R. J., (2003). *Routledge* (8th edn., p. 1). 11 New Fetter Lane, London.
6. Beer, C., Reichstein, M., Tomelleri, E., Ciais, P., Jung, M., Carvalhais, N., & Papale, D., (2010). Terrestrial gross carbon dioxide uptake: Global distribution and covariation with climate. *Science, 329*(5993), 834–838.
7. Betts, A. K., (2004). Understanding hydrometeorology using global models. *Bulletin of the American Meteorological Society, 85*(11), 1673–1688.
8. Betts, A. K., Ball, J. H., Viterbo, P., Dai, A., & Marengo, J., (2005). Hydrometeorology of the Amazon in ERA-40. *Journal of Hydrometeorology, 6*(5), 764–774.
9. Bleasdale, A., (1973). Statistical problems in practical hydrometeorology. *Journal of the Royal Statistical Society; Series D (The Statistician), 22*(2), 103–123.
10. Brenot, H., Walpersdorf, A., Reverdy, M., Van Baelen, J., Ducrocq, V., Champollion, C., & Giroux, P., (2014). A GPS network for tropospheric tomography in the framework of the Mediterranean hydrometeorological observatory Cévennes-Vivarais (southeastern France). *Atmospheric Measurement Techniques, 7*(2), 553–578.
11. Brown, J. D., & Seo, D. J., (2010). A nonparametric postprocessor for bias correction of hydrometeorological and hydrologic ensemble forecasts. *Journal of Hydrometeorology, 11*(3), 642–665.

12. Bruce, J. P., & Clark, R. H., (2013). *Introduction to Hydrometeorology: Pergamon International Library of Science, Technology, Engineering and Social Studies* (Vol. 1, pp. 1–24). Elsevier.

13. Bruijnzeel, L. A., Mulligan, M., & Scatena, F. N., (2011). Hydrometeorology of tropical montane cloud forests: emerging patterns. *Hydrological Processes, 25*(3), 465–498.

14. Brunsell, E., & Marcks, J., (2005). Identifying a baseline for teachers' astronomy content knowledge. *Astronomy Education Review, 2*(3), 38–46.

15. Byers, H. R., (1959). *General Meteorology* (Vol. 1, pp. 1–20). McGraw-Hill Book Company, New York.

16. Cao, X., Chen, J., Chen, L., Liao, A., Sun, F., Li, Y., & Peng, S., (2014). Preliminary analysis of spatiotemporal pattern of global land surface water. *Science China Earth Sciences, 57*(10), 2330–2339.

17. Central Ground Water Board (CGWB), (1991). *Bhujal News* (Vol 6, No 1). New Delhi. Special Issue on Ground Water Statistics.

18. Chen, J. L., Wilson, C. R., Chambers, D. P., Nerem, R. S., & Tapley, B. D., (1998). Seasonal global water mass budget and mean sea level variations. *Geophysical Research Letters, 25*(19), 3555–3558.

19. Cleverly, J., Chen, C., Boulain, N., Villalobos-Vega, R., Faux, R., Grant, N., & Eamus, D., (2013). Aerodynamic resistance and Penman–Monteith evapotranspiration over a seasonally two-layered canopy in semiarid central Australia. *Journal of Hydrometeorology, 14*(5), 1562–1570.

20. De Paor, D. G., Dordevic, M. M., Karabinos, P., Burgin, S., Coba, F., & Whitmeyer, S. J., (2017). Exploring the reasons for the seasons using Google Earth, 3D models, and plots. *International Journal of Digital Earth, 10*(6), 582–603.

21. Dhar, O. N., & Rakhecha, P. R., (1979). A review of rainfall relationships based upon Indian data. *Water Resources Journal* (Vol. 1, pp. 1–24). ESCAP, Bangkok, Thailand.

22. Erdil, A., & Arcaklioglu, E., (2013). The prediction of meteorological variables using artificial neural network. *Neural Computing and Applications, 22*(7), 1677–1683.

23. Eskridge, R. E., Ku, J. Y., Rao, S. T., Porter, P. S., & Zurbenko, I. G., (1997). Separating different scales of motion in time series of meteorological variables. *Bulletin of the American Meteorological Society, 78*(7), 1473–1484.

24. Flannigan, M. D., & Harrington, J. B., (1988). A study of the relation of meteorological variables to monthly provincial area burned by wildfire in Canada (1953–80). *Journal of Applied Meteorology and Climatology, 27*(4), 441–452.

25. Fletcher, R. D., (1951). Hydrometeorology in the United States. In: *Compendium of Meteorology* (Vol. 1, pp. 1033–1047). American Meteorological Society, Boston, MA.

26. Gocic, M., & Trajkovic, S., (2013). Analysis of changes in meteorological variables using Mann-Kendall and Sen's slope estimator statistical tests in Serbia. *Global and Planetary Change, 100*, 172–182.

27. Hamlet, A. F., & Lettenmaier, D. P., (2005). Production of temporally consistent gridded precipitation and temperature fields for the continental United States. *Journal of Hydrometeorology, 6*(3), 330–336.

28. Hao, L., Huang, X., Qin, M., Liu, Y., Li, W., & Sun, G., (2018). Ecohydrological processes explain urban dry island effects in a wet region, southern China. *Water Resources Research, 54*(9), 6757–6771.

29. Hess, M., Koepke, P., & Schult, I., (1998). Optical properties of aerosols and clouds: The software package OPAC. *Bulletin of the American Meteorological Society, 79*(5), 831–844.

30. Hunter, T. S., Clites, A. H., Campbell, K. B., & Gronewold, A. D., (2015). Development and application of a North American Great Lakes hydrometeorological database—Part I: Precipitation, evaporation, runoff, and air temperature. *Journal of Great Lakes Research, 41*(1), 65–77.

31. Ishii, M., Shouji, A., Sugimoto, S., & Matsumoto, T., (2005). Objective analyses of sea-surface temperature and marine meteorological variables for the 20th century using ICOADS and the Kobe collection. *International Journal of Climatology: A Journal of the Royal Meteorological Society, 25*(7), 865–879.

32. Jaybhaye, P., Raundal, P., & Andhale, P., (2013). *Handbook of Agrometeorology* (pp. 107–202). Contenental Prakashan, Vijaya Nagar, Pune – 411 030, Maharashtra, India.

33. Karyampudi, V. M., & Carlson, T. N., (1988). Analysis and numerical simulations of the Saharan air layer and its effect on easterly wave disturbances. *Journal of Atmospheric Sciences, 45*(21), 3102–3136.

34. Khavrus, V., & Shelevytsky, I., (2010). Introduction to solar motion

geometry on the basis of a simple model. *Physics Education, 45*(6), 620–641.

35. Koutsoyiannis, D., Mamassi, N., & Tegos, A., (2007). Logical and illogical exegeses of hydrometeorological phenomena in ancient Greece. *Water Science and Technology: Water Supply, 7*(1), 13–22.
36. Kramareva, L. S., Andreev, A. I., Bloshchinskiy, V. D., Kuchma, M. O., Davidenko, A. N., Pustatintsev, I. N., & Korolev, S. P., (2019). The use of neural networks in hydrometeorology problems. *Computational Technologies, 24*(6), 50–59.
37. L'vovich, M. I., (1979). In: Nace, R. L., (ed.), *World Water Resources and Their Future (English Translation from the Russian)* (Vol. 1, pp. 1–29). American Geophysical Union, Washington, D.C.
38. Li, Q., Yang, J., & Yang, L., (2021). Impact of urban roughness representation on regional hydrometeorology: An idealized study. *Journal of Geophysical Research: Atmospheres, 126*(4), e2020JD033812.
39. Lim, S., Cifelli, R., Chandrasekar, V., & Matrosov, S. Y., (2013). Precipitation classification and quantification using X-band dual-polarization weather radar: Application in the hydrometeorology testbed. *Journal of Atmospheric and Oceanic Technology, 30*(9), 2108–2120.
40. Liu, Z., Dreybrodt, W., & Wang, H., (2010). A new direction in effective accounting for the atmospheric CO2 budget: Considering the combined action of carbonate dissolution, the global water cycle and photosynthetic uptake of DIC by aquatic organisms. *Earth-Science Reviews, 99*(3–4), 162–172.
41. Marengo, J. A., (2009). Long-term trends and cycles in the hydrometeorology of the Amazon basin since the late 1920s. *Hydrological Processes: An International Journal, 23*(22), 3236–3244.
42. Milly, P. C. D., & Shmakin, A. B., (2002). Global modeling of land water and energy balances. Part II: Land-characteristic contributions to spatial variability. *Journal of Hydrometeorology, 3*(3), 301–310.
43. Nakaegawa, T., (2012). Comparison of water-related land cover types in six 1-km global land cover datasets. *Journal of Hydrometeorology, 13*(2), 649–664.
44. Neiman, P. J., Ralph, F. M., Moore, B. J., Hughes, M., Mahoney, K. M., Cordeira, J. M., & Dettinger, M. D., (2013). The landfall and

inland penetration of a flood-producing atmospheric river in Arizona. Part I: Observed synoptic-scale, orographic, and hydrometeorological characteristics. *Journal of Hydrometeorology, 14*(2), 460–484.

45. Nernec, J., (1983). The concept of runoff in the global water budget. In: *Variations in the Global Water Budget* (Vol. 1, pp. 479–488). Springer, Dordrecht.
46. Panthou, G., Mailhot, A., Laurence, E., & Talbot, G., (2014). Relationship between surface temperature and extreme rainfalls: A multi-time-scale and event-based analysis. *Journal of Hydrometeorology, 15*(5), 1999–2011.
47. Pellarin, T., Delrieu, G., Saulnier, G. M., Andrieu, H., Vignal, B., & Creutin, J. D., (2002). Hydrologic visibility of weather radar systems operating in mountainous regions: Case study for the Ardèche catchment (France). *Journal of Hydrometeorology, 3*(5), 539–555.
48. Petterssen, S., (1969). *Introduction to Meteorology* (Vol. 2, pp. 1–24). McGraw-Hill Book Company, Inc., New York.
49. Pfister, S., Bayer, P., Koehler, A., & Hellweg, S., (2011). Environmental impacts of water use in global crop production: Hotspots and trade-offs with land use. *Environmental Science and Technology, 45*(13), 5761–5768.
50. Phillips, A. D., (2012). *Principles of Flash Flood Development: An Introduction to Hydrometeorology*, *1*, 1–24.
51. Plummer, J. D., & Maynard, L., (2014). Building a learning progression for celestial motion: An exploration of students' reasoning about the seasons. *Journal of Research in Science Teaching*, *51*(7), 902–929.
52. Rakhecha, P. R., & Singh, V. P., (2009). Foreshadowing precipitation. In: *Applied Hydrometeorology* (pp. 278–299). Springer, Dordrecht.
53. Rao, K. L., (1975). *India's Water Wealth* (Vol. 1, pp. 1–24). Orient Longman, New Delhi.
54. Roy, S. B., (2011). Simulating impacts of wind farms on local hydrometeorology. *Journal of Wind Engineering and Industrial Aerodynamics, 99*(4), 491–498.
55. Sahoo, A. K., Pan, M., Troy, T. J., Vinukollu, R. K., Sheffield, J., & Wood, E. F., (2011). Reconciling the global terrestrial water budget using satellite remote sensing. *Remote Sensing of Environment, 115*(8), 1850–1865.

56. Smith, E. A., (2002). Leveraging improvements in precipitation measuring from GPM mission to achieve prediction improvements in climate. *Weather and Hydrometeorology* (Vol. 1, pp. 1–20).

57. Smithsonian Institutions, (1939). *Smithsonian Meteorology Tables* (Vol. 5, pp. 1–23). Published by the Smithsonian Institutions, Washington, D.C.

58. Snook, N., Kong, F., Brewster, K. A., Xue, M., Thomas, K. W., Supinie, T. A., & Albright, B., (2019). Evaluation of convection-permitting precipitation forecast products using WRF, NMMB, and FV3 for the 2016–17 NOAA hydrometeorology testbed flash flood and intense rainfall experiments. *Weather and Forecasting, 34*(3), 781–804.

59. Sterling, S. M., Ducharne, A., & Polcher, J., (2013). The impact of global land-cover changes on the terrestrial water cycle. *Nature Climate Change, 3*(4), 385–390.

60. Still, C. J., Berry, J. A., Collatz, G. J., & DeFries, R. S., (2003). Global distribution of C3 and C4 vegetation: carbon cycle implications. *Global Biogeochemical Cycles, 17*(1), 6–1.

61. Stohl, A., & James, P., (2004). A Lagrangian analysis of the atmospheric branch of the global water cycle. Part I: Method description, validation, and demonstration for the August 2002 flooding in central Europe. *Journal of Hydrometeorology, 5*(4), 656–678.

62. Sulis, S., Dragomir, D., Lendl, M., Bourrier, V., Demory, B. O., Fossati, L., & Winn, J. N., (2019). Multi-season optical modulation phased with the orbit of the super-Earth 55 Cancri e. *Astronomy & Astrophysics, 631*, A129.

63. Takle, E. S., Roads, J., Rockel, B., Gutowski Jr, W. J., Arritt, R. W., Meinke, I., & Zadra, A., (2007). Transferability intercomparison: An opportunity for new insight on the global water cycle and energy budget. *Bulletin of the American Meteorological Society, 88*(3), 375–384.

64. Tateishi, R., & Ahn, C. H., (1996). Mapping evapotranspiration and water balance for global land surfaces. *ISPRS Journal of Photogrammetry and Remote Sensing, 51*(4), 209–215.

65. Thornton, P. E., Running, S. W., & White, M. A., (1997). Generating surfaces of daily meteorological variables over large regions of complex terrain. *Journal of Hydrology, 190*(3–4), 214–251.

66. Trenberth, K. E., Smith, L., Qian, T., Dai, A., & Fasullo, J., (2007). Estimates of the global water budget and its annual cycle using observational and model data. *Journal of Hydrometeorology, 8*(4), 758–769.

67. Varshneya, M. C., & Balkrishana, P. P., (2018). *Text Book of Agricultural Meteorology* (pp. 1–88). ICAR, Sixth Reprint.

68. Vivoni, E. R., Gutiérrez-Jurado, H. A., Aragón, C. A., Méndez-Barroso, L. A., Rinehart, A. J., Wyckoff, R. L., & Jackson, T. J., (2007). Variation of hydrometeorological conditions along a topographic transect in northwestern Mexico during the North American monsoon. *Journal of Climate, 20*(9), 1792–1809.

69. Vörösmarty, C. J., Green, P., Salisbury, J., & Lammers, R. B., (2000). Global water resources: Vulnerability from climate change and population growth. *Science, 289*(5477), 284–288.

70. Wigley, T. M., Briffa, K. R., & Jones, P. D., (1984). On the average value of correlated time series, with applications in dendroclimatology and hydrometeorology. *Journal of Applied Meteorology and Climatology, 23*(2), 201–213.

71. Wong, S., Fetzer, E. J., Kahn, B. H., Tian, B., Lambrigtsen, B. H., & Ye, H., (2011). Closing the global water vapor budget with AIRS water vapor, MERRA reanalysis, TRMM and GPCP precipitation, and GSSTF surface evaporation. *Journal of Climate, 24*(24), 6307–6321.

72. Xavier, A. C., King, C. W., & Scanlon, B. R., (2016). Daily gridded meteorological variables in Brazil (1980–2013). *International Journal of Climatology, 36*(6), 2644–2659.

73. Xu, C. Y., & Singh, V. P., (1998). Dependence of evaporation on meteorological variables at different time-scales and intercomparison of estimation methods. *Hydrological Processes, 12*(3), 429–442.

74. Yamazaki, T., Yabuki, H., & Ohata, T., (2007). Hydrometeorological effects of intercepted snow in an eastern Siberian taiga forest using a land-surface model. *Hydrological Processes: An International Journal, 21*(9), 1148–1156.

75. Yang, J., Wang, Z. H., Georgescu, M., Chen, F., & Tewari, M., (2016). Assessing the impact of enhanced hydrological processes on urban hydrometeorology with application to two cities in contrasting climates. *Journal of Hydrometeorology, 17*(4), 1031–1047.

76. Yu, K. C., Sahami, K., Sahami, V., & Sessions, L. C., (2015). Using a digital planetarium for teaching seasons to undergraduates. *Journal of Astronomy & Earth Sciences Education, 2*(1), 33–50.

77. Zhang, J., Qi, Y., Kingsmill, D., & Howard, K., (2012). Radar-based quantitative precipitation estimation for the cool season in complex terrain: Case studies from the NOAA hydrometeorology testbed. *Journal of Hydrometeorology, 13*(6), 1836–1854.

78. Zhi, W., Feng, D., Tsai, W. P., Sterle, G., Harpold, A., Shen, C., & Li, L., (2021). From hydrometeorology to river water quality: Can a deep learning model predict dissolved oxygen at the continental scale? *Environmental Science & Technology, 55*(4), 2357–2368.

79. Ziomas, I. C., Melas, D., Zerefos, C. S., Bais, A. F., & Paliatsos, A. G., (1995). Forecasting peak pollutant levels from meteorological variables. *Atmospheric Environment, 29*(24), 3703–3711.

Chapter 2

The Hydrological Cycle

CONTENTS

2.1. INTRODUCTION

The hydrologic cycle (Figures 2.1–2.3) is a continuous sequence of moving water and transfer that occurs amongst the atmosphere, the oceans, and the earth's surface and under the earth's surface. Precipitation, Evaporation, and transpiration, groundwater, and runoff are the four essential components of the hydrologic cycle. The cycle can be viewed as starting with the water evaporation from the surface of the water. Due to the thermal energy produced by solar radiation, evaporation takes place from lakes, land surfaces, oceans, streams as well as transpiration from plants, forests, and plant leaves in the form of droplets. Water vapor flows upward, condensing, and forming clouds at greater elevations (Chahine, 1992; Held and Soden, 2006). Although the majority of the cloud settles back to the sea as precipitation while winds drive a portion of the cloud to the land. Clouds create precipitation on the ground surface in many forms, such as dew, rain, hail, snow, sleet, etc., through a variety of processes. Whenever rain falls out of the sky, some of them evaporates as it hits the ground. A portion of the precipitation that reaches the ground evaporates faster and returns to the atmosphere. The remainder penetrates within the soil, enriching its moisture content. Plants use some of the soil water, which is then transferred to the atmosphere via ET. The water remained seeps into the ground and becomes underground water. Springs, as well as other outlets, can bring groundwater to the surface. When there is a lot of rain and the soil is become saturated then water termed runoff runs across the ground. Surface runoff collects in lakes and streams before returning to the oceans. The hydrologic cycle ends in this way (Ramanathan et al., 2001; Gonçalves et al., 2011). The precipitation element of the hydrologic cycle is essential to hydrometeorologists because it shows the amount of water obtained for usage across a given area. The continuity equation also referred as the hydrologic equation or water budget equation can be used to estimate the amount of water flowing through different elements of the hydrologic cycle (Ohmura and Wild, 2002). The hydrologic equation represents the mass conservation law which is expressed as:

$$I = O + S$$

where; I is the volume of water input at a specific period; O is the volume of water outflow within the period; and S is the variation in the volume of water stored during the period. Volumes are frequently stated as average depths over just a specific area in hydrologic measurements.

It is crucial to remember that the world's total water supply is constant. The annual average evaporation out from oceans is calculated to be 3,35,000 km^3, with 65,000 km^3 from the land surface and the inland water bodies. This gives total evaporation of 4,00,000 km^3. As the average water vapor content remains unchanged and this must be equivalent to the amount of water returning to the surface in the shape of precipitation (snowfall and rainfall). We get the average yearly worldwide rainfall of 80 cm by dividing the entire evaporation volume each year by the earth's surface area ($4\pi r^2 = 4 \times 3.14 \times 6371 \times 6371 = 5.1 \times 108$ km^2).

Precipitation does not fall in a regular pattern across the globe. The zone of equatorial receives roughly 200 cm on average, whereas the zones of subtropical receive far lower than the average of 80 cm. Over the land, evaporation surpasses precipitation over through the ocean surface, and precipitation surpasses evaporation. The majority of the extra water is returned to the sea. As rain falls occur from the sky, some of it evaporates as it hits the ground. A portion of the water that reaches the ground evaporates fast and returned to the atmosphere. The rest is absorbed by the soil. Plants use some of the soil water, which is then given to the atmosphere via evapotranspiration (ET). The leftover water seeps into the ground and becomes groundwater. Slowly moving groundwater appears as streamflow or approaches the floor of the ocean. Surface water goes into lakes and rivers if rainfall is severe and the earth is saturated, ultimately reaching the seas and oceans via streamflow (Hagemann et al., 2006).

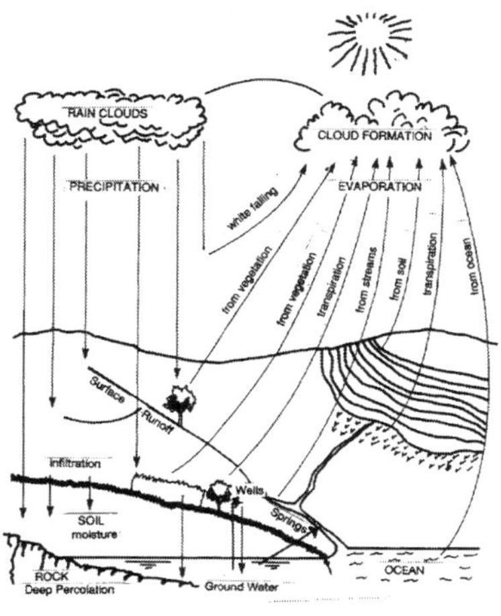

Figure 2.1. Hydrologic cycle (Ackermann et al., 1955).

Source: *https://link.springer.com/referenceworkentry/10.1007%2F1-4020-4 4 97-6_116.*

The hydrological cycle explains the water movement on, above, and beneath the Earth's surface. It is a model that explains the movement and storage of water among the atmosphere (the air enveloping the Earth and that is a gases mixture, primarily oxygen (20%)and nitrogen (80%) with some minor gases), the biosphere (all ecosystems global sum, sometimes referred to as the life zone on Earth crust), the anthroposphere (human effect on the Earth system), the hydrosphere, the lithosphere (the Earth outermost rigid shell consist of a part of the uppermost mantle and the crust), and the cryosphere (the places of ice and snow) (see Table 2.1).

Soil-vegetation-atmosphere transfer schemes (SVATs) or land surface parameterization schemes (LSPs) are terms used to describe biosphere models. Sellers et al. (1986) provided an example of an SVAT. In addition to the sea, water can be found on the Earth's surface in the form of wetlands, streams, and lakes. The water of the surface also contains ice and snow, which are solid types of precipitation. Groundwater is the water that exists under the Earth's surface (Koutsoyiannis, 2020).

The majority of energy left the surface of the ocean as latent heat in water vapor, still, this is not always true for land surfaces. As a result, air masses of maritime differ from air masses of continental. The precipitation, transfer of energy, momentum at their interface, and water vapor are all-important ties between the oceans and the atmosphere. The oceans, which are denser as compare to the atmosphere and have higher mechanical inertia, offer a massive reservoir of stored energy. As a result, currents of the ocean move far more slowly than the currents of the atmosphere. Energy from the sun is collected by the subsurface and heats the atmosphere from below, while the oceans are heated from above. In continental locations, underground water, lakes, rivers can have major hydro climatological and hydrometeorological importance. The simplified graphic in Figure 2.1 depicts the hydrological cycle (Mook and Rozanski, 2000; Bullock and Acreman, 2003).

Table 2.1. Distribution of Freshwater on the Earth and Water in the Hydrosphere (Martinec, 1985)

(a) Distribution of Water in the Hydrosphere (Baumgartner and Reichel, 1975)		
Forms of Water Present	**Water Volume (10^6 km^3)**	**Percentage of Volume (%)**
Oceans, seas	1348	97.4
Polar ice, sea ice, glaciers	28	2.0
Surface water, groundwater, atmospheric water	8	0.6

Total	**1384**		**100**	
Total Freshwater	**36**		**2.6**	
(b) Distribution of Fresh Water on Earth (Volker, 1970; Dracos, 1980)				
Forms of Water Present	**Water Volume (10⁶ km³)**		**Percentage of Volume (%)**	
	*	†	*	†
Polar ice, glaciers	24.8	27.9	76.93	77.24
Soil moisture	0.09	0.06	0.28	0.17
Groundwater within reach	3.6	3.56	11.17	9.85
Deep groundwater	3.6	4.46	11.17	12.35
Lakes and rivers	0.132	0.127	0.41	0.35
Atmosphere	0.014	0.014	0.04	0.04
Total	**32.236**	**36.121**	**100.0**	**100.0**

*Based on Volker (1970)

†Based on Dracos (1980)

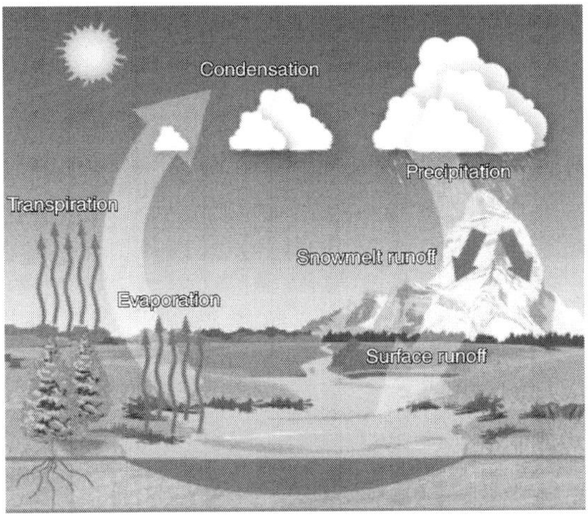

Figure 2.2. Hydrological cycle simplified representation.

Source: *NWS Jetstream NOAA, USA, www.srh.noaa.gov/jetstream/atmos/hydro.htm*

Water is constantly circulated between the reservoirs. C condensation, deposition, precipitation, runoff, sublimation, infiltration, groundwater flow, transpiration, melting, and evaporation all contribute to this cycling. The probable residence durations of water in the main reservoirs are shown in Table 2.2. In rivers, water is replenished every 16 days on average. Every eight days, the water in the atmosphere is entirely replenished. Groundwater, ocean bodies, glaciers, and large lakes have lower rates of replenishment. It can take hundreds of years for these reservoirs to be replenished. The Nubian aquifer, located below the Sahara Desert (Egypt), contains some of the world's oldest groundwater with water ages ranging from 200 ka to 1 million years ago based on the area (Ohmura and Wild, 2002).

Humans are depleting a few of these assets (particularly groundwater) at rates that are significantly faster than they can be replenished. This form of resource consumption effectively makes the water non-renewability (Kahn and Uhlenbeck, 1938). Every year, 577,000 km^3 of water is exchanged on the surface of the Earth. This is the amount of water that evaporates from the land (74,200 km^3) and the surface of the ocean (502,800 km^3). 119,000 km^3 on of waterfall on the land and 458,000 km^3 on the ocean and an equal amount as atmospheric precipitation. Overall runoff of the rivers (42,700 km^3/year) and direct runoff of groundwater to the ocean (2,100 km^3/year) are represented by the difference among evaporation and transpiration from the surface of the land (119,000–74,200 = 44,800 km^3/year). These are the primary sources of clean water for man's daily needs and commercial activity (Wilson, 1974; Schaefer and Bloomfield, 2004).

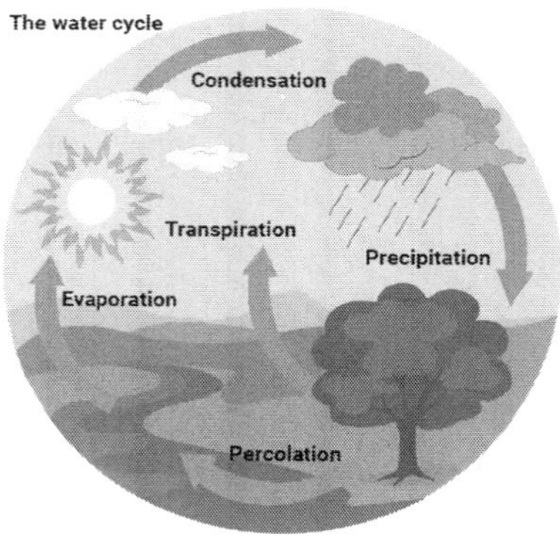

Figure 2.3. The hydrological cycle is depicted in a simplified diagram.

Source: *https://knowitinfo.com/what-is-water-cycle-in-short/.*

2.2. INFLOWS AND OUTFLOWS

The water cycle, also known as the hydrologic cycle, is a natural mechanism that recycles the water supply of e The sun provides energy to this cycle, which runs continually (Ehlin, 1981; Zhang and Mitsch, 2005). Storage, inflows, and outflows are all part of the hydrologic cycle. Inflows provide water to various sections of the hydrologic system, and outflows take it away. The retaining of water by components of the system is referred to as storage. An influx for one portion of the system is an efflux for the other since water has cyclical movement (Klemeš, 1983; Sutcliffe and Parks, 1987).

In the case of an aquifer, water percolation into the ground is an aquifer's intake. An outflow is the groundwater discharge from an aquifer into a stream (also stream inflow). If aquifer inflows exceed outflows, the amount of stored water in the aquifer will rise over time. In contrast, if the aquifer's inflows are smaller than its outflows, the number of stored water declines. Outflows and Inflows can happen by themselves or as a result of human action (Hunt et al., 2006; Davis, 2008).

2.2.1. Condensation

The cooling process of water vapor till it forms a liquid is known as condensation. Water vapor condenses as the dew point is achieved, forming small visible droplets of water. Clouds occur when such droplets develop in the sky as well as various atmospheric conditions exist. When the droplets meet, they mix to build big droplets, resulting in precipitation (Jaqaman et al., 1983; Heath and Rock, 2002).

Hot air rises cool and loses its ability to contain water vapor, resulting in condensation in the atmosphere. The excess vapor of water condenses to generate cloud droplets. Unstable air convection, raising of air by fronts and raising over higher topography like mountains, and convergence linked with cyclones can all create vertical motions that cause clouds (Langer, 2000; Jones, 2004).

2.2.2. Precipitation

Hail, sleet, snow, or rain, are all examples of moisture falling from the atmosphere. Geographic locations and seasons affect the intensity, amount, and precipitation type. These variables influence whether the water flows towards streams or seeps into the earth. Rainfall and snow records are preserved in most places of the world. This enables scientists to calculate average rainfall for a given site and categorize rainstorms according to average return period, duration, and severity. This data is essential for sustainable agriculture, flood control, and the design of water control structures (Huffman et al., 1997; Trenberth et al., 2003).

Water vapor can condense back, releasing latent heat that is transformed into sensible heat, warming the surrounding air. This warmth encourages airlifting, which aids in condensation and adiabatic cooling. As water droplets coalesce into bigger droplets, they grow large enough to drop as precipitation on the surface of the earth. Because they are high in the troposphere, they have a lot more potential energy, which is transformed to kinetic energy as they fall down the surface. They turn this kinetic energy to work performed on the surface (for example, erosion) by hitting it. Water vapor may be moved to another area in which it condenses and precipitates out instead of precipitating and condensing back into the ocean (New et al., 2001; Stephens et al., 2010).

2.2.3. Evaporation

The phase shift of water into vapor is known as evaporation. Evaporation is a primary way for energy to be transferred from the surface to the air. "Latent energy" is the energy being required to evaporate water. As water transitions from a liquid state to a gas occur so the latent energy is "imprisoned" within the water molecule. Around 60° south and 60° north latitude, the ocean accounts for 88% of all water reaching the atmosphere. The majority of the water that evaporates from the ocean goes back to the ocean. Before being precipitated out, some water is moved over land (Penman, 1956a, b; Monteith, 1965). The process of turning water from a liquid state to a gas state. The oceans account for about 80% of total evaporation, with the other 20% originating from vegetation and inland water. Winds distribute evaporated water across the world, affecting the air humidity everywhere.

Even outside of clouds, most water evaporates remains as a gas, and evaporation is much more severe when temperatures are higher. The highest evaporation was happening across the oceans and toward the equator (Thom and Oliver, 1977; Granger and Gray, 1989). The simplified description in details of transpiration process and affecting elements are given in FAO-56 (Allen et al., 2006) as:

1. **Evaporation:** It is the process whereby liquid water is converted to water vapor (vaporization) and removed from the evaporating surface (vapor removal). Water evaporates from a variety of surfaces, such as lakes, rivers, pavements, soils, and wet vegetation. And for this process energy is required to change the state of the molecules of water from liquid to vapor. The energy gets for this process either direct from solar radiation and, to a

lesser extent, the ambient temperature. While, the driving force to remove water vapor from the evaporating surface is the difference between the water vapor pressure at the evaporating surface and that of the surrounding atmosphere.

The elements involved in the processes of evaporation are:

i. Climatological parameters of evaporation process:

a. Solar radiation;

b. Air temperature;

c. Air humidity; and

d. Wind speed.

ii. Other factors that affect the evaporation process: Where the evaporating surface is the soil surface, the degree of shading of the crop canopy and the amount of water available at the evaporating surface.

Wet Soil: Where the soil is able to supply water fast enough to satisfy the evaporation demand, the evaporation from the soil is determined only by the meteorological conditions.

Dry Soil Surface: Evaporation decreases rapidly and may cease almost completely within a few days.

2. Transpiration: It consists of the vaporization of liquid water contained in plant tissues and the vapor removal to the atmosphere. Whereas, crops predominately lose their water through stomata and these are small openings on the plant leaf through which gases and water vapor pass. The vaporization occurs within the leaf, namely in the intercellular spaces, and the vapor exchange with the atmosphere is controlled by the stomatal aperture (Figure 2.4). Transpiration, like direct evaporation, depends on the energy supply, vapor pressure gradient and wind. Hence, radiation, air temperature, air humidity and wind terms should be considered when assessing transpiration.

Transpiration depends on the different factors viz., energy supply, vapor pressure gradient and wind. Whereas, climatological parameters of transpiration process were radiation, air temperature, air humidity and wind terms. While, other factors that affect the transpiration process are the soil water content and the ability of the soil to conduct water to the roots also determine the transpiration rate, as do waterlogging and soil water salinity.

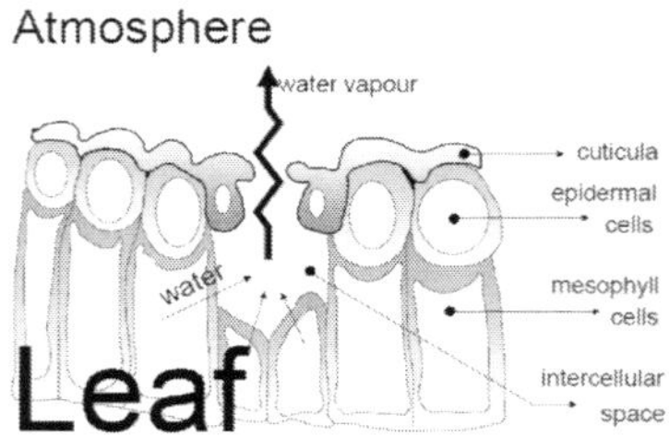

Figure 2.4. Schematic representation of stomata and escaping water molecules from leaf by transpiration.

Source: *FAO-56.*

And the transpiration rate is also influenced by crop characteristics, environmental aspects, and cultivation practices. Different kinds of plants may have different transpiration rates. Not only the type of crop, but also the crop development, environment, and management should be considered when assessing transpiration.

2.2.4. Evapotranspiration (ET)

The net effect of transpiration and evaporation is called evapotranspiration (ET). ET consumes a greater proportion of precipitation than other hydrologic cycle processes. The returning process of moisture to the atmosphere is known as evaporation. The sun's heat warms water on every surface, particularly the surfaces of oceans, mudholes, streams, lakes, rivers, and ponds until it approaches the point where it changes into gaseous, or vapor form. After then, the water vapor rises to the atmosphere (Hargreaves and Samani, 1982, 1985).

The concept of ET is explained in short and simply in FAO-56 (Allen et al., 2006); the combination of two separate processes whereby water is lost on the one hand from the soil surface by evaporation and on the other hand from the crop by transpiration is referred to as ET. Evaporation and transpiration occur simultaneously and there is no easy way of distinguishing between the two processes. Evaporation is the process whereby liquid water is converted to water vapor (vaporization) and removed from the evaporating surface (vapor removal). Water evaporates from a variety of surfaces, such as lakes, rivers, pavements, soils, and wet vegetation. Energy is required to change the state of the molecules of water from liquid to vapor. Direct solar radiation and, to a lesser extent, the ambient temperature. The driving force to remove water vapor from the evaporating surface is the difference between the water vapor pressure at the evaporating surface and that of the surrounding atmosphere.

1. **Units of ET:** The ET rate is normally expressed in millimeters (mm) per unit time. The rate expresses the amount of water lost from a cropped surface in units of water depth. The time unit can be an hour, day, decade, month or even an entire growing period or year. As one hectare has a surface of 10 000 m^2 and 1 mm is equal to 0.001 m, a loss of 1 mm of water corresponds to a loss of 10 m^3 of water per hectare. In other words, 1 mm day^{-1} is equivalent to 10 m^3 ha^{-1} day^{-1}. However, the factors affecting on the ET are weather, soil/topography and crop factors, and which is illustrated well in Figure 2.5.
2. **Weather Parameters:** Factors affecting on ET: The principal weather parameters affecting ET are radiation, air temperature, humidity, and wind speed.
3. **Crop Factors:** The crop type, variety, and development stage of crop should be considered when assessing the ET from crops grown in large, well-managed fields. Differences in resistance to transpiration, crop height, crop roughness, reflection, ground cover and crop rooting characteristics result in different ET levels in different types of crops under identical environmental conditions.
4. **Other Factors Soil/Topography:** Influencing other elements on ET are ground cover, plant density and the soil water content. The effect of soil water content on ET is conditioned primarily by the magnitude of the water deficit and the type of soil. On the other hand, too much water will result in water logging which might damage the root and limit root water uptake by inhibiting respiration.

Several procedures have been developed to assess the evaporation rate from these parameters. The evaporation power of the atmosphere is expressed by the reference crop evapotranspiration (ET_0 or PET). The reference crop ET represents the ET from a standardized vegetated surface. Crop ET under standard conditions (ET_c) refers to the evaporating demand from crops that are grown in large fields under optimum soil water, excellent management, and environmental conditions, and achieve full production under the given climatic conditions. Crop ET under nonstandard conditions ($ET_{c\ adj}$) is the ET from crops grown under management and environmental conditions that differ from the standard conditions. When cultivating crops in fields, the real crop ET may deviate from ET_c due to non-optimal conditions such as the presence of pests and diseases, soil salinity, low soil fertility, water shortage or waterlogging. This may result in scanty plant growth, low plant density and may reduce the ET rate below ET_c.

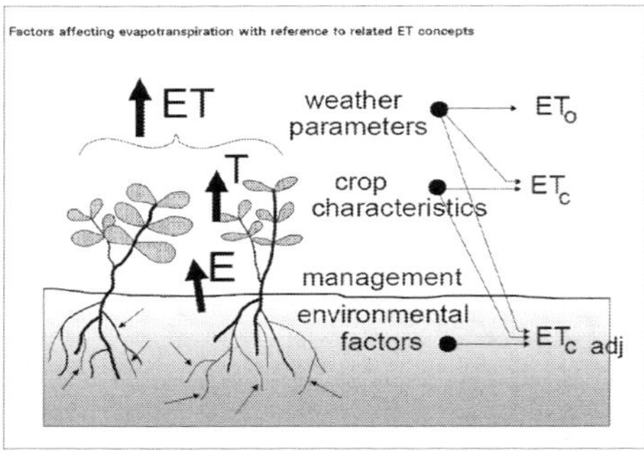

Figure 2.5. Schematic representation of factors affecting on ET with reference to related ET concepts.

Source: *FAO-56.*

The process through which plants transfer moisture into the air is known as transpiration. Water is taken up by plants using their roots and some part of it is lost via pores present in their leaves. The water moisture in the leaves takes energy and evaporates into the air as warm air flows across them. The water evaporation from the stems and leaves of plants into the atmosphere is known as transpiration.

Soil water is absorbed by plants through its roots, which can come from quite deep within the soil. (Corn plants, for example, has roots that reach 2.5 meters into the earth, whereas some desert plants have roots that reach 20 meters). Plants use water to give nutrients to their leaves by pumping them up from the soil. The water evaporation through microscopic pores called "stomates" located on the lower part of leaves drives this pumping. Transpiration is responsible for around 10% of all water evaporation (Kristensen and Jensen, 1975).

Transpiration and evaporation both release water into the atmosphere, which is referred to as ET. The amount of water released to the air by these processes is known as actual ET. Actual ET is a water outflow that is influenced by humidity, surface moisture, and temperature. Consider ET to be a form of water consumption, in which water is transpiring and evaporation as a result of which environmental conditions of locations are define. As long as there are plants transpire and water evaporates, actual ET rises as temperatures rise. ET is also influenced by the available amount of water, which is determined by soil field capacity (Pereira et al., 1999).

2.2.5. Infiltration and Percolation

The penetration of water into the surface of the soil is known as infiltration. Infiltration is the only source of water for the development of vegetation, and it also serves to keep the water supply of ground to springs, streams going, and wells. The physical properties of the soil, rainfall intensity, soil water content, soil cover (i.e., plants), and soil temperature all affect the infiltration rate. Percolation and Infiltration are phrases that are frequently used identically (Wangemann et al., 2000).

Water is captured by plants or drops directly on the surface as it approaches the surface in various kinds of precipitation. Interception is the term for precipitation that accumulates on the stems or leaves of plants. The quantity of water a plant intercept is greatly determined by its shape. Water is retained on the surface of the leaf until it either trickles down or falls the leaf stem, eventually approaching the ground via stem flow. The surface is protected from degradation by intercepting falling rain. Because deciduous trees lose their leaves for some time, coniferous trees capture higher water every year than deciduous trees (Hur et al., 2009).

A certain amount of water infiltrates through the soil after approaching the ground perhaps seeping down to the zone of groundwater, or running as runoff across the surface. Water that seeps through the soil's surface is referred to as infiltration. Soil moisture state, soil structure, soil texture, and vegetation all influence infiltration. In dry soils, significant infiltration rates occur, but as the soil becomes wetter, infiltration slows. Infiltration rates are higher in coarse-textured soils with well-connected, large pore spaces as compared to small textured soils. Due to the relatively low amount of pore space within a soil unit volume, coarse-textured soils load quicker than small textured soils. Runoff is produced faster than it would be with a smaller grained soil.

Infiltration is also influenced by vegetation. Infiltration is faster in soils with forest vegetation than in bare soils. Tree roots loosen the soil and offer channels for water to enter. The impact of incoming rain is reduced by surface litter and foliage, which prevents soil passageways from being clogged (Figure 2.6).

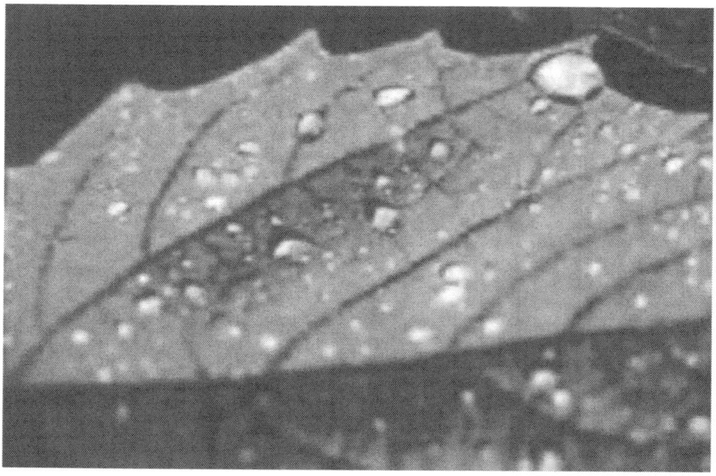

Figure 2.6. Water droplets intercepted by a tree leaf.

Source: *https://www.sciencedirect.com/topics/earth-and-planetary-sciences/hydrological-cycle.*

The downward passage of water via rock and soil is known as percolation. Percolation takes place below the root zone. Groundwater percolates into the soil like water via a sponge, moving from one location to another over sand, gravel, rock, or tunnels in formations such as cavernous limestone. Percolation and Infiltration are phrases that are frequently used interchangeably. The quantity of water kept in the soil at any given time is referred to as Moisture Storage in the Soil. Soil factors such as organic matter content and soil texture influence the water amount in the soil. The field capacity is the maximum water quantity that the soil can retain. Field capacities are higher in fine grain soils than in coarse grain soils. As a result, fine soils have more water accessible for genuine ET over coarse (sandy) soils. soil moisture storage is found within field capacity and 0. And the quantity of water that is given to or withdrawn from what is retained in the soil is referred to as the Soil Moisture Storage Changes. The variation in soil moisture storage occurs within field capacity and 0.

2.2.6. Runoff

Runoff is the flow of water all over the planet's crust, generally from oceans, precipitation, lakes, stream channels, or low points or depressions (Göbel et al., 2007). Rainfall frequency and duration, as well as soil type, the slope of the ground, and ground cover, all influence the rate of runoff. Precipitation can either fall directly to the earth or be apprehended by plants before approaching it. Water can penetrate the soil or flow out across the ground as runoff when it hits the ground. Surface runoff takes place when rainfall intensity surpasses infiltration capacity or when the soil's water retention capacity is reached. Soil texture and structure influence infiltration and water retention. Because sand-rich soils have significant, well-connected pore spaces, water can infiltrate them quickly. Because of their tiny pore spaces, clay soils have low infiltration rates. In a unit volume of coarse, sandy soil, however, there is less total pore space than that in clay-rich soil. As a consequence, sandy soils fill quickly and produce runoff much more quickly, unlike clay soils (Pinder and Jones, 1969).

Surface runoff occurs when annual rainfall surpasses the soil's infiltration rate, or when the soil's field capacity has been achieved. Water is either constrained or unconstrained as it flows throughout the surface. Sheet erosion occurs when uncontrolled flow moves around the surface in broadsheets. Water is restricted to channels in bounded flow. A version of bounded flow is streamflow.

Water that flows along the surface can get locked in depressions and be stored there. As the depression fills, water may either evaporate back into the atmosphere, penetrate the soil surface, or leak out.

Stream discharge is one method of measuring streamflow. The volume of water that passes through a specific cross-section of a stream in a given amount of time is known as stream discharge. The volume of water flowing through a stream is measured in cubic feet per second or cubic meters per second. The leakage of groundwater into the stream channel provides the stream's "normal" or base flow. This drainage is what allows perennially flowing streams to remain active throughout the year. When a storm produces precipitation, the stream discharge rises as more water is added to the stream, either directly into the channel or through runoff (Hsieh and Davis, 2005). The flow of land water to the oceans, primarily in the form of rivers, streams, and lakes, is known as runoff. Precipitation that does not evaporate, transpire, or penetrate the surface to become groundwater is known as runoff. Even the tiniest streams are linked to larger bodies of water that hold billions of gallons of water to the world's oceans. Flooding can happen as a consequence of excessive runoff, which takes place because there is too much rainfall (Martínez-Vilalta et al., 2014).

2.2.7. Transport

The water flow through the atmosphere, particularly from the ocean waters to the earth, is referred to as transportation. Clouds, which are made up of tiny water dewdrops and/or ice crystals, carry a few of the earth's water evaporation. The jet stream, surface-based circulations like land and sea winds, and other processes accelerate clouds from one location to another. A typical 1-kilometer dense cloud, on the other hand, only holds adequate water for a millimeter of rainfall, whilst the amount of moisture in the air is typically 10–50 times higher (Passioura, 1988; Kalra et al., 2003). Water vapor, which is the third most plentiful gas in the atmosphere, transports the majority of the water. Water vapor may be unseen to us, but it is observable to satellites that can measure the relative humidity (RH) of the air. Visualizations like this water vapor image are created using this data, giving a visual representation of moisture absorption in the air (Van den Honert, 1948; McCulloh et al., 2003).

2.3. THE HYDROLOGICAL CYCLE COMPRISING PROCESSES

The hydrological cycle involves several procedures, the most vital of which are as described in the following (Thornes, 1999; Gouveia et al., 2019):

1. **Transpiration:** The evaporation of water from plants occurs via tiny holes on the underside of leaves (known as stomata). Most plants' transpiration is a passive activity governed largely by atmospheric humidity and soil water content. Only 1% of the water transpired by a plant is used to grow, with the remaining being released into the air. Evaporation and transpiration revert backwater to the atmosphere at different rates depending on the climate (Da Silva et al., 2021).
2. **Evaporation:** The transition from liquid to vapor is known as evaporation. The Earth, the atmosphere, the Sun, or human activity may provide the necessary energy.
3. **Infiltration:** If the ground is not saturated or has fissures, water will infiltrate the land surface. Water may drift into the ground

and recharge aquifers, or it may move via unsaturated zones and discharge into the sea, rivers, and lakes. Streamflow is maintained between storms or snowmelt periods by the outflow from groundwater sources. When storms are severe, the majority of the water quickly approaches streams. Overland flow can occur if the water level-the demarcation between the saturated and unsaturated zones-soars to the earth's surface (Storm and Refsgaard, 1990).

4. **Condensation:** It is the transformation of atmospheric water vapor into liquid water in the form of clouds and dew. This is dependent on the temperature of the dew point and the temperature of the air. The dew point temperature is the point at which the air becomes saturated and dew can form as it cools. Water vapor condenses if the temperature is lowered further. Mist and fog form when the ambient temperature and the dew point temperature are both the same. Because water vapor has a higher energy density than liquid water, surplus energy is emitted in the form of heat when condensation occurs. When tiny condensation molecules collide or coalesce to the point where the elevated air can no longer support them, precipitation forms on the Earth's surface. Freshwater strikes the surface primarily through precipitation, which averages 980 mm per year over both oceans and land.

5. **Residence Time:** The average time a molecule of water spends in a specific area is the residence time of water in different parts of the water cycle. It is worth mentioning that groundwater can spend over 10,000 years beneath the Land crust before disappearing, whilst water preserved in the soil only lasts a few days. But when the water has evaporated, it spends about nine days in the surrounding air before condensing and falling to the Earth's crust as precipitation. There are two methods for estimating residence times. The first and most common approach is to apply the conservation of mass principle, which assumes that the proportion of water in a given warehouse is relatively constant. Divide the amount of water in the warehouse by the rate at which water enters or leaves the store to get the retention time. Isotropic methodologies are the second process for groundwater. Instream tracer injection coupled with modeling or measurements of naturally produced tracers like radon 222 are used in these methods (Lamontagne and Cook, 2007).

Small particles (aerosols) are discharged into the air due to human activities, which may increase radiation from the sun's scattering and absorption. They also result in brighter clouds which are less effective at releasing moisture. These aerosol impacts can result in a feeble hydrological cycle, which is directly linked to potable water availability and quality (Ramanathan et al., 2001).

Table 2.2. Average Residence Period for Specific Stores

Reservoir	Average Residence Time
Antarctica	20,000 years
Oceans	3,200 years
Glaciers	20 to 100 years
Seasonal snow cover	2 to 6 months
Soil moisture	1 to 2 months
Groundwater shallow	100 to 200 years
Groundwater deep	10,000 years
Lakes	50 to 100 years
Rivers	2 to 6 months
Atmosphere	9 days

2.4. GLOBAL IMPACTS ON THE HYDROLOGICAL CYCLE

The main reason for the general movement of the atmosphere is differential heating by the Solar system. Aside from the Sun's and Earth's relative positions, the hydrological cycle is influenced by several regional variations. The earth's rotation causes latitudinal disparities in solar input. Weather patterns in circulation patterns result, as does a frequent diurnal cycle in longitude (Bala et al., 2008, 2011).

In both continental and oceanic surfaces, the flows of superficial sensible and latent heat vary. In addition, the topography of continents diverges in ways that have a substantial impact on the atmosphere. Changes in land cover, sea surface temperature, and atmospheric composition have a long-term and short-term impact on the water cycle (D'Odorico et al., 2019).

Clouds move large amounts of water through the atmosphere. As a result, they have a significant effect on solar radiation absorption and alter the surface energy intake. A few erect clouds shadow the ground and prevent solar radiation from reaching it, whilst high clouds prevent high-frequency radiation from escaping (Held and Soden, 2006; Myhre et al., 2018).

For both the overall average and the mean overseas (Figure 2.7(c)), the peak zonal mean precipitable water is found at 10°N (Figure 2.7(a)). The annual mean of zonal mean precipitable water over land (Figure 2.7(b)) is nearly equal around the equator, and the summer (JJA) mean is nearly identical to the winter (DJF) mean between 40°S and 40°N. Through the saturation vapor pressure, the water vapor content of the atmosphere is highly sensitive to heat, and temperature decreases as one moves higher in the atmosphere. The pascal (Pa) is the SI unit of pressure, and atmospheric pressure at sea level is about 105 Pa (1,000 hectopascals) (hPa). More than half of the water vapor is concentrated underneath 850 hPa, and over 90% is restrained to the interface underneath 500 hPa (Peixoto and Oort, 1992).

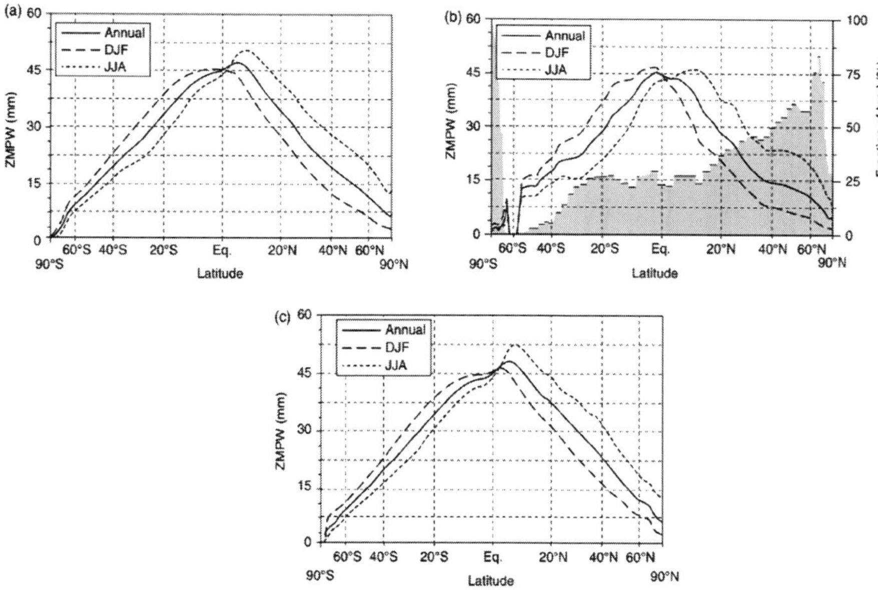

Figure 2.7. The zonal mean precipitable water distribution along the meridian (mm): (a) over sea and land; (b) mean over land only; and (c) means overseas only. Yearly mean, December-January-February (DJF) mean, and June-July-August (JJA) mean, from 1989 to 1992 (Oki, 1999).

Source: *https://www.amazon.com/Hydrometeorology-Advancing-Weather-Climate-Science/dp/1118414977.*

2.5. WATER BALANCE

Ground analyses like precipitation, water storage in lakes, and groundwater have typically been used to evaluate the water cycle at the earth's surface. This is defined as follows (Milly, 1994; Xu and Singh, 1998):

$$\frac{\partial S}{\partial t} = -\nabla_H \cdot R_0 - \nabla_H \cdot R_u - (E - P) \quad (1)$$

where; H denotes horizontal divergence; S denotes water storage in the area; P denotes precipitation; $\mathbf{R}_u$ denotes groundwater mobility; $\mathbf{R}_0$ denotes runoff water; and E denotes ET. Snow cover, as well as surface water, soil moisture, and groundwater storage, including retention water, are all represented by S (Oki, 1999).

The water balance in the atmosphere is mentioned by:

$$\frac{\partial W}{\partial t} + \frac{\partial W_c}{\partial t} = -\nabla_H \cdot Q - \nabla_H \cdot Q_c - (E - P) \quad (2)$$

where; W_c denotes solid and liquid water column storage; W denotes precipitable water in vapor in a column; $\mathbf{Q}_c$ denotes vertically integrated 2-dimensional water flux in solid and liquid phases; $\mathbf{Q}$ denotes vertically integrated 2-dimensional water vapor flux. Figure 2.8 summarizes both the earth's surface and atmospheric water balances (Belmans et al., 1983; Stephenson, 1990).

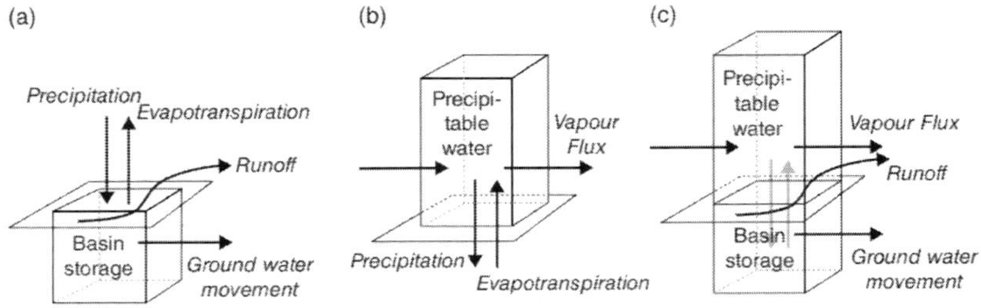

Figure 2.8. (a) Water balance of terrestrial; (b) water balance of atmosphere; and (c) combined terrestrial-atmospheric water balance (Oki, 1999).

Source: *https://www.amazon.com/Hydrometeorology-Advancing-Weather-Climate-Science/dp/1118414977.*

2.6. AEROSOLS IMPACT ON THE HYDROLOGICAL CYCLE

Aerosol particles such as solid and liquid particulate matter apart from natural droplets of water play a significant role in the water cycle because they serve as condensation nuclei for cloud droplet formation and disperse and absorb radiation from the sun (Douville et al., 2002; Boughton, 2004).

This has been described by Ramanathan et al. (2001). Aerosols in the atmosphere can be as small as a 100th of a micrometer or as large as tens of micrometers. They come from a variety of places, including the Earth's surface, volcanoes, biosphere, and oceans. Many aerosols are produced as a result of human behavior. Huge quantities of aerosol are produced by massive quantities of dust from significant deserts, like the Sahara. The largest natural flux of aerosol mass into the atmosphere is sea salt, which is generated by collapsing air bubbles and winds, even so, most of this mass is large particles that are not traveled far off. Large volumes of gases and particles can be injected into the surrounding air by volcanoes, and if they approach the stratosphere, they are taken away more gradually than if they stay in the troposphere. Sulfates are formed from sulfur gases released by fossil fuel combustion, continental biota, and marine phytoplankton, are a significant contributor of aerosols (Lamarque et al., 2005).

Rises in cloud condensation nuclei (CCN), which are aerosols larger than a few hundredths of a micrometer in size, boost cloud microbubble number concentration, resulting in increased cloud extent and global cooling. If the water availability for condensation in the cloud remains constant, there will be tinier drops, that are less expected to expand large enough to result in precipitation (see Chapter 2); consequently, clouds will last longer, contributing to cooling once again. The hydrological cycle will be weakened by the freezing of the surface resulting in increased solar energy backscattering as well as the lower productivity of clouds in making precipitation due to higher aerosol (Ramanathan et al., 2005; Takahashi et al., 2018).

2.7. HYDROLOGICAL CYCLE COUPLED MODELS

Reviews between hydrological, ecological, and atmospheric systems, along with human society, are frequently investigated in hydrological cycle research. Unpredicted responses are frequently the result of system feedback. As a result, numerical modeling experts face difficulties in coupling multiple sections of the Earth system (Rahmstorf, 1995; Oleson et al., 2008). It is not possible to draw a line between the elements of the simulated geo-ydro-biosphere system. It is impossible to model findings of two or more combined state variables without also modeling the procedures that link them.

Integrating unlike Earth system components inside of hydrological models has progressed dramatically over the last 20 years (see for instance Bronstert et al., 2005). Hydrology, it is clear, must interrelate with not only atmospheric sciences, but also geochemistry, sociology, soil sciences, and even biology (Hack et al., 2006).

Scale considerations are crucial. At large levels, methods that influence the behavior of the system at small levels may become insignificant, and new procedures may arise as essential to the scale change. Furthermore, the spatial variability of fluxes like rainfall, characteristics like hydraulic conductivity, and state parameters like soil humidity frequently results in unanticipated system responses. The persistence of drier or wetter climatic conditions is also influenced by sub-surface and land surface circumstances, especially soil moisture. This is significant in the event of large floods or droughts. Eventually, natural, and human reactions are crucially significant (Hagemann et al., 2006).

2.8. WATER CYCLE EXCHANGES AND GLOBAL ENERGY PROJECT (GEWEX)

GEWEX is a multi-faceted science initiative, observation, and research. It tends to focus on hydrologic, terrestrial, atmospheric, and radiative. Energy and radiation transitions, global, and regional hydrological cycles, and their role in climate change are all determined by interactions and coupled processes. It is the centerpiece of the WCRP (World Climate Research Program), which is devoted to the study of atmospheric dynamics and thermodynamics. It also studies interactions with the Planet's crust, as well as its implications for the global energy and water cycles (Chahine, 1992; Coughlan and Avissar, 1996).

The purpose of GEWEX is to recreate and anticipate variations in the global hydrological cycle by using appropriate numerical frameworks. It affects surface and atmospheric dynamics, as well as variations in regional water resources and hydrological processes, besides their responses to environmental changes like increased greenhouse gas emissions (GHG).

GEWEX will enhance the effectiveness to design global rainfall and evaporation by an order of magnitude, furthermore provide an adequate evaluation of atmospheric radiation and cloud sensitivity to climatic changes (Trenberth and Asrar, 2012).

GEWEX is an important part of the WCRP's engagement with many international organizations and programs that cope up with climate findings. GEWEX contributes its technical strengths in two particular societal benefit domains: water and climate, as part of WCRP's insight to the group on earth observations (GEO) global earth observation system of systems (GEOSS). GEWEX is the driving force behind the worldwide data reprocessing activity and observation plan. It uses GEOSS interpretative platforms as a feasibility study for future climate change comprehension. GEWEX is a supporter of the IGWCO (integrated global water cycle observations) theme (van Oevelen, 2015).

Around 70% of the Earth's surface is covered by clouds. The GEWEX Cloud Assessment, which began in 2005, is the world's first coordinated intercomparison of freely accessible global cloud products derived from lidar measurements, multi-spectral imagers, and IR sounders (Gash et al., 2001).

2.9. FLOODING

Flooding occurs when water covers land that is not usually covered by it. It can occur from a variety of sources, including the following (Blom and Voesenek, 1996; Christensen and Christensen, 2003):

1. **Coastal Flooding:** Sea levels can rise due to tidal and weather circumstances. Flooding of this type is expected to become more common and severe in the future.
2. **Groundwater Flooding:** When rainfall affects naturally collected underground aquifers to rise to the surface, low-lying regions can be flooded.
3. **River (Fluvial) Flooding:** When a river cannot handle the amount of water flowing into it, this happens. Over 12% of the UK population lives in fluvial floodplains or coastal flood-prone areas. River flooding affects more than 150 million people and $1,300 billion in global GDP every year.
4. **Drain and Sewer (Urban) Flooding:** When drains are blocked or overflowing, this can happen during heavy rain.

5. **Surface Water (Pluvial) Flooding:** This occurs when there is heavy rain on already saturated ground or covered areas with poor drainage.

Flooding's potential impacts appeared to have increased between 1970 and 2010 (Jongman et al., 2012). In 2010, Asia was home to three-quarters of those severely impacted by river flooding. Only 9% of people lived in Europe, although, after Asia, Europe had the highest number of people affected by river flooding. Between now and 2050, the population of Europe subjected to river flooding is expected to decrease. Flooding caused $19 billion in global financial losses in 2012, and the number is steadily increasing (Nelson and Pope, 1978; Ward et al., 2013).

Flooding is one of the major effects of climate change, and it has the potential to cause significant loss of life and property in many regions of the world. The throughput of 11 different climate models was used in a recent study to evaluate the enhanced global flood risk due to climate change (Hirabayashi et al., 2013). They were capable of predicting that floods will become much more recurrent over 42% of the Planet's crust using a river routing method to assess future river flows and flooding zones. The northern Andes, India, South East Asia, and eastern Africa are expected to be the hardest hit. Floods that used to happen once every 100 years are expected to happen every 5–25 years in many places (Emmelkamp and Wessels, 1975; Bailey-Serres and Voesenek, 2008).

REFERENCES

1. Allen, R. G., Pereira, L. S., Raes, D., & Smith, M., (2006). *Crop Evapotranspiration (Guidelines for Computing Crop Water Requirements)*. Irrigation and Drainage Paper No. 56, PDF version. FAO, Water Resources, Development and Management Service, Rome, Italy.
2. Bailey-Serres, J., & Voesenek, L. A. C. J., (2008). Flooding stress: Acclimations and genetic diversity. *Annu. Rev. Plant Biol.*, *59*, 313–339.
3. Bala, G., Caldeira, K., Nemani, R., Cao, L., Ban-Weiss, G., & Shin, H. J., (2011). Albedo enhancement of marine clouds to counteract global warming: impacts on the hydrological cycle. *Climate Dynamics, 37*(5), 915–931.
4. Bala, G., Duffy, P. B., & Taylor, K. E., (2008). Impact of geoengineering schemes on the global hydrological cycle. *Proceedings of the National Academy of Sciences*, *105*(22), 7664–7669.
5. Baumgartner, A., & Reichel, E., (1975). *World Water Balance* (Vol. 1, pp. 1–23). Oldenbourg, Munich.
6. Belmans, C., Wesseling, J. G., & Feddes, R. A., (1983). Simulation model of the water balance of a cropped soil: SWATRE. *Journal of Hydrology*, *63*(3–4), 271–286.
7. Blom, C. W. P. M., & Voesenek, L. A. C. J., (1996). Flooding: the survival strategies of plants. *Trends in Ecology & Evolution*, *11*(7), 290–295.
8. Boughton, W., (2004). The Australian water balance model. *Environmental Modelling & Software*, *19*(10), 943–956.
9. Bullock, A., & Acreman, M., (2003). The role of wetlands in the hydrological cycle. *Hydrology and Earth System Sciences*, *7*(3), 358–389.
10. Chahine, M. T., (1992). GEWEX: The global energy and water cycle experiment. *Eos, Transactions American Geophysical Union*, *73*(2), 9–14.
11. Chahine, M. T., (1992). The hydrological cycle and its influence on climate. *Nature*, *359*(6394), 373–380.
12. Christensen, J. H., & Christensen, O. B., (2003). Severe summertime flooding in Europe. *Nature*, *421*(6925), 805–806.

13. Coughlan, M., & Avissar, R., (1996). The global energy and water cycle experiment (GEWEX) continental-scale international project (GCIP): An overview. *Journal of Geophysical Research: Atmospheres*, *101*(D3), 7139–7147.
14. D'Odorico, P., Carr, J., Dalin, C., Dell'Angelo, J., Konar, M., Laio, F., & Tuninetti, M., (2019). Global virtual water trade and the hydrological cycle: Patterns, drivers, and socio-environmental impacts. *Environmental Research Letters*, *14*(5), 053001.
15. Da Silva, H., Barbosa, A. P. D., Silva, L. N. L., Oliveira Junior, E. S., Corrêa, F., Kantek, D. L. Z., & Muniz, C. C., (2021). Ichthyofauna in an ecological station in the Pantanal of Brazil: The hydrological cycle affecting species composition. *Arxius de Miscel· lània Zoològica*, *18*, 183–193.
16. Davis, A. P., (2008). Field performance of bioretention: Hydrology impacts. *Journal of Hydrologic Engineering*, *13*(2), 90–95.
17. Douville, H., Chauvin, F., Planton, S., Royer, J. F., Salas-Melia, D., & Tyteca, S., (2002). Sensitivity of the hydrological cycle to increasing amounts of greenhouse gases and aerosols. *Climate Dynamics*, *20*(1), 45–68.
18. Dracos, Th., (1980). *Hydrology* (Vol. 1, pp. 1–20). Springer, Wien.
19. Ehlin, U., (1981). Hydrology of the Baltic Sea. In *Elsevier Oceanography Series* (Vol. 30, pp. 123–134). Elsevier.
20. Emmelkamp, P. M., & Wessels, H., (1975). Flooding in imagination vs flooding *in vivo*: A comparison with agoraphobics. *Behavior Research and Therapy*, *13*(1), 7–15.
21. Gash, J., Hoff, H., Kabat, P., Pitman, A., Try, P., & Seelig, W., (2001). Biospheric aspects of the hydrological cycle (BAHC) and global energy and water cycle experiment (GEWEX) news. *Global Energy and Water Cycle Experiment (GEWEX) News*, *11*(2), 1–22.
22. Göbel, P., Dierkes, C., & Coldewey, W. G., (2007). Storm water runoff concentration matrix for urban areas. *Journal of Contaminant Hydrology*, *91*(1–2), 26–42.
23. Gonçalves, H. C., Mercante, M. A., & Santos, E. T., (2011). Hydrological cycle. *Brazilian Journal of Biology*, *71*(1), 241–253.
24. Gouveia, N. A., Gherardi, D. F. M., & Aragão, L. E. O. C., (2019). The role of the Amazon River plume on the intensification of the hydrological cycle. *Geophysical Research Letters*, *46*(21), 12221–12229.

25. Granger, R. J., & Gray, D. M., (1989). Evaporation from natural nonsaturated surfaces. *Journal of Hydrology, 111*(1–4), 21–29.
26. Hack, J. J., Caron, J. M., Yeager, S. G., Oleson, K. W., Holland, M. M., Truesdale, J. E., & Rasch, P. J., (2006). Simulation of the global hydrological cycle in the CCSM community atmosphere model version 3 (CAM3): Mean features. *Journal of Climate, 19*(11), 2199–2221.
27. Hagemann, S., Arpe, K., & Roeckner, E., (2006). Evaluation of the hydrological cycle in the ECHAM5 model. *Journal of Climate, 19*(16), 3810–3827.
28. Hargreaves, G. H., & Samani, Z. A., (1982). Estimating potential evapotranspiration. *Journal of the Irrigation and Drainage Division, 108*(3), 225–230.
29. Hargreaves, G. H., & Samani, Z. A., (1985). Reference crop evapotranspiration from temperature. *Applied Engineering in Agriculture, 1*(2), 96–99.
30. Heath, R. J., & Rock, C. O., (2002). The Claisen condensation in biology. *Natural Product Reports, 19*(5), 581–596.
31. Held, I. M., & Soden, B. J., (2006). Robust responses of the hydrological cycle to global warming. *Journal of Climate, 19*(21), 5686–5699.
32. Hsieh, C. H., & Davis, A. P., (2005). Evaluation and optimization of bioretention media for treatment of urban storm water runoff. *Journal of Environmental Engineering, 131*(11), 1521–1531.
33. Huffman, G. J., Adler, R. F., Arkin, P., Chang, A., Ferraro, R., Gruber, A., & Schneider, U., (1997). The global precipitation climatology project (GPCP) combined precipitation dataset. *Bulletin of the American Meteorological Society, 78*(1), 5–20.
34. Hunt, W. F., Jarrett, A. R., Smith, J. T., & Sharkey, L. J., (2006). Evaluating bioretention hydrology and nutrient removal at three field sites in North Carolina. *Journal of Irrigation and Drainage Engineering, 132*(6), 600–608.
35. Hur, S. O., Jung, K. H., Sonn, Y. K., Ha, S. K., Kim, J. G., & Kim, N. W., (2009). Classification of hydrologic soil groups of soil originated from limestone by assessing the rates of infiltration and percolation. *Korean Journal of Soil Science and Fertilizer, 42*(2), 103–109.
36. Jaqaman, H., Mekjian, A. Z., & Zamick, L., (1983). Nuclear condensation. *Physical Review C, 27*(6), 2782.

37. Jones, G., (2004). The K Noevenagel Condensation. *Organic Reactions*, *15*, 204–599.
38. Kahn, B., & Uhlenbeck, G. E., (1938). On the theory of condensation. *Physica*, *5*(5), 399–416.
39. Kalra, A., Garde, S., & Hummer, G., (2003). Osmotic water transport through carbon nanotube membranes. *Proceedings of the National Academy of Sciences*, *100*(18), 10175–10180.
40. Klemeš, V., (1983). Conceptualization and scale in hydrology. *Journal of Hydrology*, *65*(1–3), 1–23.
41. Koutsoyiannis, D., (2020). Revisiting the global hydrological cycle: is it intensifying? *Hydrology and Earth System Sciences*, *24*(8), 3899–3932.
42. Kristensen, K. J., & Jensen, S. E., (1975). A model for estimating actual evapotranspiration from potential evapotranspiration. *Hydrology Research*, *6*(3), 170–188.
43. Lamarque, J. F., Kiehl, J. T., Hess, P. G., Collins, W. D., Emmons, L. K., Ginoux, P., & Tie, X. X., (2005). Response of a coupled chemistry-climate model to changes in aerosol emissions: Global impact on the hydrological cycle and the tropospheric burdens of OH, ozone, and NOx. *Geophysical Research Letters*, *32*(16), 1–34.
44. Lamontagne, S., & Cook, P. G., (2007). Estimation of hyporheic water residence time *in situ* using ^{222}Rn disequilibrium. *Limnol. Oceanogr. Methods*, 5, 407–416.
45. Langer, J. S., (2000). Theory of the condensation point. *Annals of Physics*, *281*(1–2), 941–990.
46. Martinec, J., (1985). Time in hydrology; Chapter 9. In: Rodda, J. C., (ed.), *Facets of Hydrology* (Vol. II, pp. 249–290). Wiley, Chichester.
47. Martínez-Vilalta, J., Poyatos, R., Aguadé, D., Retana, J., & Mencuccini, M., (2014). A new look at water transport regulation in plants. *New Phytologist*, *204*(1), 105–115.
48. McCulloh, K. A., Sperry, J. S., & Adler, F. R., (2003). Water transport in plants obeys Murray's law. *Nature*, *421*(6926), 939–942.
49. Milly, P. C. D., (1994). Climate, soil water storage, and the average annual water balance. *Water Resources Research*, *30*(7), 2143–2156.
50. Monteith, J. L., (1965). Evaporation and environment. In: *Symposia of the Society for Experimental Biology* (Vol. 19, pp. 205–234). Cambridge University Press (CUP) Cambridge.

51. Mook, W. G., & Rozanski, K., (2000). Environmental isotopes in the hydrological cycle. *Principles and Applications* (Vols. I, IV, and V).

52. Myhre, G., Samset, B. H., Hodnebrog, Ø., Andrews, T., Boucher, O., Faluvegi, G., & Voulgarakis, A., (2018). Sensible heat has significantly affected the global hydrological cycle over the historical period. *Nature Communications*, *9*(1), 1–9.

53. Nelson, R. C., & Pope, G. A., (1978). Phase relationships in chemical flooding. *Society of Petroleum Engineers Journal*, *18*(05), 325–338.

54. New, M., Todd, M., Hulme, M., & Jones, P., (2001). Precipitation measurements and trends in the twentieth century. *International Journal of Climatology: A Journal of the Royal Meteorological Society*, *21*(15), 1889–1922.

55. Ohmura, A., & Wild, M., (2002). Is the hydrological cycle accelerating? *Science*, *298*(5597), 1345–1346.

56. Oki, T., (1999). The global water cycle. Chapter 1.2. In: Browning, K. A., & Gurney, R. J., (eds.), *Global Energy and Water Cycles* (Vol. 1, pp. 10–29). Cambridge University Press, Cambridge.

57. Oleson, K. W., Niu, G. Y., Yang, Z. L., Lawrence, D. M., Thornton, P. E., Lawrence, P. J., & Qian, T., (2008). Improvements to the community land model and their impact on the hydrological cycle. *Journal of Geophysical Research: Biogeosciences*, *113*(G1), 1–23.

58. Passioura, J. B., (1988). Water transport in and to roots. *Annual Review of Plant Physiology and Plant Molecular Biology*, *39*(1), 245–265.

59. Peixoto, J. P., & Oort, A. H., (1992). *Physics of Climate* (Vol. 1, pp. 1–23). American Institute of Physics.

60. Penman, H. L., (1956a). Estimating evaporation. *Eos, Transactions American Geophysical Union*, *37*(1), 43–50.

61. Penman, H. L., (1956b). Evaporation: An introductory survey. *NJAS Wageningen Journal of life Sciences*, *4*(1), 9–29.

62. Pereira, L. S., Perrier, A., Allen, R. G., & Alves, I., (1999). Evapotranspiration: Concepts and future trends. *Journal of Irrigation and Drainage Engineering*, *125*(2), 45–51.

63. Pinder, G. F., & Jones, J. F., (1969). Determination of the ground-water component of peak discharge from the chemistry of total runoff. *Water Resources Research*, *5*(2), 438–445.

64. Rahmstorf, S., (1995). Bifurcations of the Atlantic thermohaline circulation in response to changes in the hydrological cycle. *Nature*, *378*(6553), 145–149.

65. Ramanathan, V. C. P. J., Crutzen, P. J., Kiehl, J. T., & Rosenfeld, D., (2001). Aerosols, climate, and the hydrological cycle. *Science*, *294*(5549), 2119–2124.
66. Ramanathan, V., Chung, C., Kim, D., Bettge, T., Buja, L., Kiehl, J. T., & Wild, M., (2005). Atmospheric brown clouds: Impacts on South Asian climate and hydrological cycle. *Proceedings of the National Academy of Sciences*, *102*(15), 5326–5333.
67. Ramanathan, V., Cutzen, P. J., Kiehl, J. T., & Rosenfeld, D., (2001) Aerosols, climate, and the hydrological cycle. *Science*, *294*(7), 2119–2124.
68. Schaefer, J. P., & Bloomfield, J. J., (2004). The Dieckmann Condensation (Including the T Horpe-Z Iegler condensation). *Organic Reactions*, *15*, 1–203.
69. Sellers, P. I., Mintz, Y., Sud, Y. C., & Dalcher, A., (1986). A simple biosphere model (SiB) for use within general circulation models. *J. Atmos. Sci.*, 43, 505–531.
70. Stephens, G. L., L'Ecuyer, T., Forbes, R., Gettelmen, A., Golaz, J. C., Bodas-Salcedo, A., & Haynes, J., (2010). Dreary state of precipitation in global models. *Journal of Geophysical Research: Atmospheres*, *115*(D24).
71. Stephenson, N. L., (1990). Climatic control of vegetation distribution: The role of the water balance. *The American Naturalist*, *135*(5), 649–670.
72. Storm, B., & Refsgaard, A., (1990). Distributed physically-based modelling of the entire land phase of the hydrological cycle. In: *Distributed Hydrological Modelling* (pp. 55–69). Springer, Dordrecht.
73. Sutcliffe, J. V., & Parks, Y. P., (1987). Hydrological modelling of the Sudd and Jonglei Canal. *Hydrological Sciences Journal*, *32*(2), 143–159.
74. Takahashi, H. G., Watanabe, S., Nakata, M., & Takemura, T., (2018). Response of the atmospheric hydrological cycle over the tropical Asian monsoon regions to anthropogenic aerosols and its seasonality. *Progress in Earth and Planetary Science*, *5*(1), 1–14.
75. Thom, A. S., & Oliver, H. R., (1977). On Penman's equation for estimating regional evaporation. *Quarterly Journal of the Royal Meteorological Society*, *103*(436), 345–357.

76. Thornes, J. B., (1999). The hydrological cycle and the role of water in Mediterranean environments. In: *Rural Planning from an Environmental Systems Perspective* (Vol. 1, pp. 85–107). Springer, New York, NY.

77. Trenberth, K. E., & Asrar, G. R., (2012). Challenges and opportunities in water cycle research: WCRP contributions. *The Earth's Hydrological Cycle*, 1, 515–532.

78. Trenberth, K. E., Dai, A., Rasmussen, R. M., & Parsons, D. B., (2003). The changing character of precipitation. *Bulletin of the American Meteorological Society*, *84*(9), 1205–1218.

79. Van den Honert, T. H., (1948). Water transport in plants as a catenary process. *Discussions of the Faraday Society*, *3*, 146–153.

80. Van Oevelen, P. J., (2015). Remote sensing applications in water resources and the global energy and water exchanges project. In: *AGU Fall Meeting Abstracts* (Vol. 2015, pp. H51T-04).

81. Wangemann, S. G., Kohl, R. A., & Molumeli, P. A., (2000). Infiltration and percolation influenced by antecedent soil water content and air entrapment. *Transactions of the ASAE*, *43*(6), 1517.

82. Ward, P. J., Jongman, B., Sperna, W. F., Bouwman, A. A., Van Beek, R., Bierkens, M. F. P., Ligtvoet, W., & Winsemius, H. C., (2013). Assessing flood risk at the global scale: Model setup, results, and sensitivity. *Env. Res. Lett.*, *8*, 044019.

83. Wilson, E. L., (1974). The static condensation algorithm. *International Journal for Numerical Methods in Engineering*, *8*(1), 198–203.

84. Xu, C. Y., & Singh, V. P., (1998). A review on monthly water balance models for water resources investigations. *Water Resources Management*, *12*(1), 20–50.

85. Zhang, L., & Mitsch, W. J., (2005). Modelling hydrological processes in created freshwater wetlands: An integrated system approach. *Environmental Modelling and Software*, *20*(7), 935–946.

Chapter 3

Fundamentals of Precipitation and its Changing Characteristics

CONTENTS

3.1. INTRODUCTION

Precipitation is a product produced as a result of condensation water vapors present in atmosphere that drop because of gravitational force from the clouds (Karger et al., 2017). Precipitation comprises different forms such as rain, snow, hail, sleet, hail, pellets of ice, graupel, and drizzling. When a specific atmospheric area is highly saturated with water vapors Precipitation takes place. Causing water to condense and "precipitate" or fall. Hence, fog as well as mist is colloids instead of precipitates, since the water vapors do not condense adequately to produce precipitation. Two processes probably perform together; can result in saturated air: either cooling of air or by introducing water in the form of vapors to air. Precipitation appears in the form of small droplets mixed through collision with rain drops or crystals of ice or inside a cloud. Further, small, and severe spells of rain in dispersed localities are known as showers (Braham, 1964; Ralph et al., 2018).

Moisture specifically lifted or else enforced to move over sub-freezing layer of the atmosphere at the surface could be condensed into rain and clouds.

This method is usually active while freezing rain befalls. A motionless front often exists close to the freezing rain area and obliges as the focus to rise and force air. As long as there is essential and adequate moisture content in air, the moisture inside the rising air condenses to form clouds, specifically cumulonimbus and nimbostratus in case of substantial precipitation. Ultimately, the cloud droplets will enlarge sufficient to create raindrops and fall away towards the Earth and freeze upon interaction with bare objects. If comparatively cozy water bodies are extant, for instance, as water is evaporated from lakes, lake outcome snowfall develops a downwind of the warm-lakes around the rear into the cold cyclonic flow of more hot cyclones.

Lake effect snowfall can be more in the vicinity. Within a cyclone's comma head and lake-effect precipitation bands, thunder-type snow is likely. In mountain areas, more precipitation is probable where upslope flow is exploited within windward sides of the terrain at height.

On the protected mountains side, desert climates can occur because of the dry air triggered via compressional heating. Maximum rain happens inside the tropics and is produced through convection (Manor-Chapman, 2013; Adler et al., 2019). Rainy seasons are brought to savannah regions by the migration of the monsoon trough, or intertropical convergence zone (Anagnostou, 2004).

Rain is an important aspect of the water cycle because it is responsible for replenishing the earth's water supply. Annual precipitation totals around 505,000 cubic kilometers, with 398,000 cubic kilometers falling over oceans and 107,000 cubic kilometers falling on land.

Known the Earth's surface area, that worth the globally averaged annual precipitation is 990 mm (39.6 inch), but overland, it is merely 715 mm (28.6 inch). Classification of climate systems for instance the Köppen climate classification system uses average yearly rainfall to support distinguish among various climate regimes (Graves et al., 2008; Abbott et al., 2019).

Precipitation might ensue on further celestial bodies. Such as, Saturn's major satellite, Titan, hosts precipitation of methane gas like a slow-falling shower, and witnessed as rain puddles in the Polar and equator regions (Turtle et al., 2011; Dhingra et al., 2021).

3.2. PRECIPITATION MECHANISMS

The water cycle (Figure 3.1) is incomplete without precipitation. It is liable to deposit major quantity of new water on the earth. Almost 5.05 lkm^3 of evaporated water falls as precipitation every year over the earth, out of which, 3.98 lkm^3 of water falls above the oceans. Provided the surface area of earth, which shows the globally averaged yearly precipitation is around 99.0 cm.

Precipitation mechanisms comprise orographic rainfall, convective, and stratiform as described by Dore et al. (2006). Convective methods implicate strong vertical motions which result in the upending of the atmosphere in that locality inside an hour and produce heavy precipitation, whereas stratiform procedures encompass less intense precipitation and weaker upward motions (Mapes and Houze, 1995; Pearce, 2002).

Generally, precipitation is distributed into three types, depending upon if it falls in the form of liquid water, which freezes after contacts with the surface, or ice. However, combinations of various kinds of precipitation, comprising different kinds, can fall instantaneously.

Rain and drizzle are the liquid forms of rain. Rain or drizzle that freezes on interaction with a subfreezing air mass is known as freezing drizzle or freezing rain. Frozen types of rain comprise snow, graupel, ice pellets, hail, and ice needles (Johnson and Shepherd, 2018).

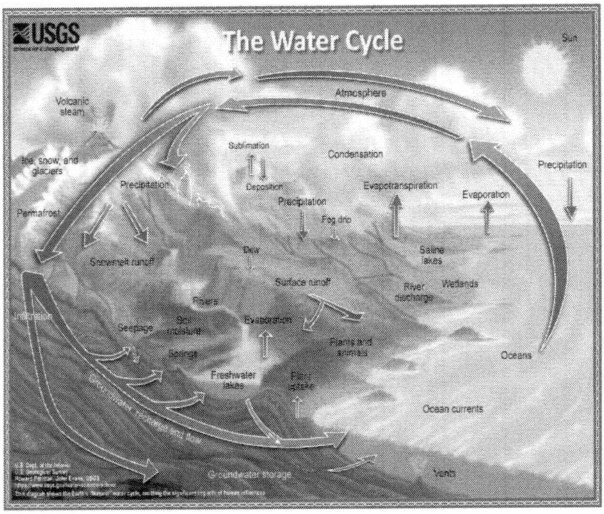

Figure 3.1. The water cycle comprising condensation and coalescence.

Source: *https://study.com/academy/lesson/the-water-cycle-precipitation-condensation-and-evaporation.html.*

3.3. PRECIPITATION MEASUREMENT

3.3.1. Measurement of Liquid Precipitation

A rain gauge is used to measure the rainfall (comprising drizzle or rain) in millimeters (mm), which corresponds to a kg/m^2. This is equal to the L/m^2 if it is assumed that 01 liter of water is equal to 1 kg of mass, which is conventional for all practical purposes. It is less expressed in centimeters. The conforming English unit employed is inches. In Australia earlier than metrication, rainfall has been calculated in "points" called hundredth of an inch (Mossop and Carte, 1957; Pidwirny, 2008).

3.3.2. Measurement of Solid Precipitation

Snow gauge is ordinarily utilized to calculate the solid rain quantity. Snowfall is generally calculated in cm via allowing snowfall into a vessel and then measures the height. The snow is melted to acquire the equivalent water measurement in millimeters as in case of liquid precipitation. The correlation among equivalent water and snow height the is determined by its water content; the water equivalent thus provides a rough approximation of depth of snow. Additional types of solid precipitation comprising even sleet, hail or snow pellets can also be melted to measure the water equal in mm as stated by (Businger and Reed, 1989; Shry and Reiley, 2010).

3.4. MECHANISM OF AIR SATURATION

3.4.1. Dew Point

The temperature at which a piece of air becomes saturated and condenses to water is known as the dew point. Water vapors usually condense on condensation nuclei like salt, dust, and ice to procedure clouds. An eminent portion of an anterior zone enforces extensive areas of lift, creating cloud decks like cirrostratus or altostratus. Moreover, stratus is termed as steady cloud deck which inclines to procedure when a steady and cool mass of air is confined beneath a mass of warm air. It also forms because of the lifting of advection fog for a period of breezy conditions (Figure 3.2) (Lee et al., 1997; Glickman and Zenk, 2000).

Figure 3.2. The late-summer rainstorm in Denmark.

Source: *https://en.wikiversity.org/wiki/Radiation_astronomy/Hydrometeors.*

Four key mechanisms exist to cool the air about its dew point: conductive cooling, adiabatic cooling, evaporative cooling, and radiational cooling. Conductive cooling takes place when the air interacts with a relatively cold surface, ordinarily propelled from one surface to another surface. Adiabatic cooling takes place when the air upsurges and expansion occurs (Braun, 2010; Patel and Cox, 2017). Air rise occur might be because of atmospheric motions on a huge scale, convection, or a physical barricade like a mountain (Figure 3.3). Radiational cooling is attributed to the release of IR by the surface beneath or by air. Evaporative cooling happens when moisture content in air is increased due to evaporation, that compels the air temperature to decrease to its temperature of wet-bulb, or till capacity (Cavallo and Hakim, 2009; Saunders et al., 2018).

Figure 3.3. The lenticular cloud formation due to mountains over Wyoming.

Source: https://www.pinterest.com/pin/243194448601157880/.

3.4.2. Increase in Moisture Content of Air

The foremost ways through which water vapors content in air is increased comprise convergence of wind into areas of ascending motion, virga falling or precipitation, transpiration from plants, daylight heating vaporizing water from oceans, wetland, moving dry air over warmer water or cool, and thrilling air over mountains (Wakimoto, 1985; Ferguson and Wood, 2011).

3.5. TYPES OF PRECIPITATION

3.5.1. Rainfall

Coalescence arises when water precipitations merge to form bigger water precipitations, or freeze on top of an ice crystal, that is called the Bergeron method. The rate of falling of tiny precipitations is insignificant, therefore clouds of these tiny droplets do not fall to the earth; rain will merely arise when these droplets merge into bigger water drops. When the air turbulence ensues, these precipitations collide with one another, creating larger water droplets (Figure 3.4). When these larger precipitations incline, coalescence lasts, so that larger drops develop sufficiently heavy to overwhelmed resistance of air and result in rain (Laws, 1941).

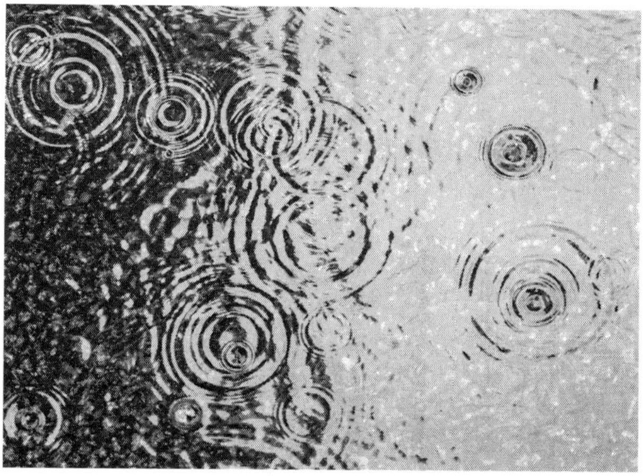

Figure 3.4. The formation of puddle in the rain.

Source: *https://theconversation.com/what-happens-when-a-raindrop-hits-a-puddle-115984.*

Raindrops bear sizes ranging from 0.1 millimeters to 9 millimeters mean diameter, beyond that they are susceptible to breakdown. Minor drops are entitled cloud precipitations having a spherical shape. As a raindrop rises in magnitude, its shape develops to more pumpkin-shaped, with its biggest cross-section facing the approaching airflow. In contrast to the cartoon images of raindrops, their outline does not look like a teardrop (Oguntoyinbo and Akintola, 1983; Houze, 1997).

Extent and intensity of rainfall are generally inversely proportional. Therefore, storms of high-intensity are expected to have little duration whereas low-intensity storms might have extended duration. Raindrops related to melting hail have a propensity for a larger size than other raindrops. RA is the METAR code for rain, whereas the code for rain dips stands for SHRA (Conway et al., 2004; Junker, 2008).

3.5.2. Ice Pellets or Sleet Formation

Sleet, often known as ice pellets, is a type of rain that consists of microscopic, luminous ice balls. These are frequently lower in size than sleets. They bounce a lot when they reach the ground, and most ice pellets do not freeze into a solid mass because they do not want to mix with the freezing rain (Rutledge et al., 2006).

Ice pellets are formed when an air that is above freezing layer subsists along with sub-freezing air layer. This originates from the fractional or widespread melting of snowflakes dropping across the warm layer. When they contingency into the layer of sub-freezing nearer to the surface, they re-freeze to form sleet or ice pellets. Though, if the sub-freezing layer is considerably small underneath the warm layer, the precipitation will freeze again to form ice pellets in a short time interval, and freezing rain will be the consequence at the surface. Presentation of a temperature profile in the cold season, a warm layer above the ground is more likely to be initiate ahead of a heartfelt front, but it can also be located overdue a transient cold front (Figure 3.5) (Rauber et al., 2001).

Figure 3.5. Accumulated ice pellets.

Source: *https://en.wikipedia.org/wiki/Ice_pellets.*

3.5.3. Hail Formation

Similar to former precipitation, hail is formed when supercooled water droplets in storm clouds freeze after having interaction with the condensation nuclei. The condensation nuclei are particles in air such as dust particles or dirt. The hailstones are blown by the updraft of the storm towards the higher portion of the cloud. The updraft dispels and the sleets start to fall under the updraft, and are then lifted over. Generally, hail possesses a diameter of almost 5 millimeters (0.20 in) or in some bags greater than 5 millimeters Precipitation of small pieces of ice, with diameter ranging from 5 to 50 mm or something more, is known as hail and hailstorm is the most dreaded and destructive form of precipitation produced during thunderstorms or from cumulonimbus clouds. Hail is a solid, frozen form of precipitation that causes extensive damage to properties and growing crops. Hot, humid afternoon hours during the summer are the most congenial for development of hailstorms. These usually form over a relatively small area and pass over within a very short period, however, causes havoc even from a few minutes spell (Mesinger and Mesinger, 1992; Menon, 1993; Jewell and Brimelow, 2009). GR is the METAR code used to specify larger hail having diameter greater than 6.4 millimeters. The abbreviation GR comes from the French word grêle. However, GS is used for the small-sized hail and snow pellets, derived from the French word grésil. Moreover, hailstones bigger than golf ball are frequently reported hail sizes (Lozowski et al., 1991). Hailstones could develop to 15 centimeters (6 in) in diameter and may attain the weight of more than 500 grams (1 lb). Outer shell of the large hailstone could melt because of release of latent heat by additional freezing. The hailstone endures 'wet growth,' and resultantly the melted outer shell gathers further smaller hailstones. Later, the hailstone achieves a layer of ice and develops gradually larger with every ascent until it becomes excessively heavy to be sustained by the storm's updraft, and it sprays from the cloud (Figure 3.6) (Brimelow et al., 2002; Vivoda et al., 2018).

Figure 3.6. A big hailstone having diameter of 6 centimeters (2.4 in).

Source: *https://en.wikipedia.org/wiki/Hail.*

3.5.3.1. Hail Characteristics and Damage Potential

Usually, the extent of damage by hail is determined by the size and density of hailstones that fall per unit area and wind force during the hail fall event. There are three types of hail classified according to the occurrence and nature; soft hail (less than 5 mm ice mixture when falling on the earth surface breaks into small pieces is called Soft hail); small hail (small ice crystals falling on the earth along with rainfall but do not break down into small pieces is called small hail); severe hail (Hard hails having diameter of 5.5 micrometer commonly known as Severe hail) and according size, shape, and nature hailstones are classified into different classes with photographs and its damaging potential are explained (Rao et al., 2014).

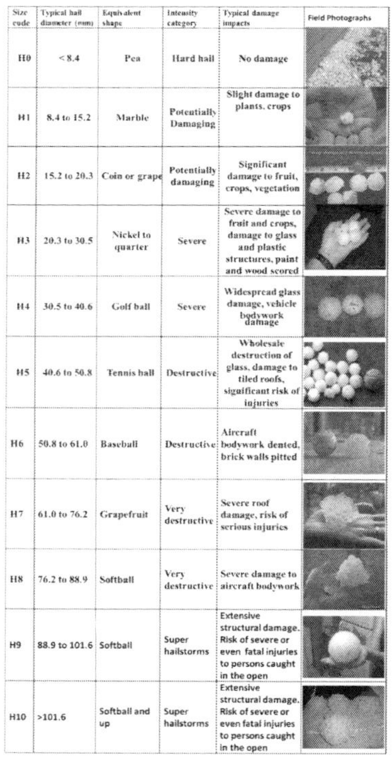

Size code	Typical hail diameter (mm)	Equivalent shape	Intensity category	Typical damage impacts	Field Photographs
H0	<8.4	Pea	Hard hail	No damage	
H1	8.4 to 15.2	Marble	Potentially Damaging	Slight damage to plants, crops	
H2	15.2 to 20.3	Coin or grape	Potentially damaging	Significant damage to fruit, crops, vegetation	
H3	20.3 to 30.5	Nickel to quarter	Severe	Severe damage to fruit and crops, damage to glass and plastic structures, paint and wood scored	
H4	30.5 to 40.6	Golf ball	Severe	Widespread glass damage, vehicle bodywork damage	
H5	40.6 to 50.8	Tennis ball	Destructive	Wholesale destruction of glass, damage to tiled roofs, significant risk of injuries	
H6	50.8 to 61.0	Baseball	Destructive	Aircraft bodywork dented, brick walls pitted	
H7	61.0 to 76.2	Grapefruit	Very destructive	Severe roof damage, risk of serious injuries	
H8	76.2 to 88.9	Softball	Very destructive	Severe damage to aircraft bodywork	
H9	88.9 to 101.6	Softball	Super hailstorms	Extensive structural damage. Risk of severe or even fatal injuries to persons caught in the open	
H10	>101.6	Softball and up	Super hailstorms	Extensive structural damage. Risk of severe or even fatal injuries to persons caught in the open	

Figure 3.7. Pictorial description of hailstone intensity scale and nature of damage.

Source: *Rao et al. (2014); Photos from collection of Dr. Pralhad Jaybhaye; Maharashtra farmers face impacts of hailstorms and state's "Inaction" Plan on Climate Change| SANDRP South Asia Network on Dams, Rivers, and People (SANDRP), 7/5/2017.*

The damage with hails is determined by the size ranges and the number of hailstones that fall per unit area during a hail fall, wind force during the event and the property of the target. The extent of crop-hail damages also varies depending on the stage of occurrence of hail during the crop growing season (Figure 3.7). While, the different crops are damaged in varying ways by hail.

In India, an unprecedented 20 days was prevailed in different states, during February 24 to March, 20 14, has left millions of farmers in a state of shock; surprisingly in 2014, almost 28 out of 35 districts in Maharashtra were severely affected by unseasonal rains and heavy hailstorm causing widespread damage to the agriculture and other infrastructures. According to the elders, in most of the affected villages, this is the first life time experience of heavy damage from hailstorm in Maharashtra during last 60 years (Anonymous, 2014). Usually, hailstorm activity occurs during the months of April and May. But this time it has occurred during February-March which is a severe weather aberration. Over 1.24 lakh ha. of farmland spread over 1086 Villages in Maharashtra have been adversely affected in the on-going spell of unseasonal rain and hailstorm (February 10–13, 2018) which started on 10 February (Biswas and Marar, 2018).

Figure 3.8. Nature of hailstones/beds/sheets on different terrains of Baramati, Beed, and Wasim districts of Maharashtra state (India) was observed during 24 February, to end of March, 2014.

Source: *Photos from collection of Dr. Pralhad Jaybhaye; South Asia Network on Dams, Rivers, and People (SANDRP), 7/5/2017.*

The comprehensively recent worst affected area by hailstorm events in India were reported by different organizations, media, and personals (Anonymous, 2014, 2018; SANDRP, 2017). And actual pictures of real time hailstorm events picked and collected through multimedia by Dr. Pralhad Jaybhaye are showed in Figures 3.8–3.11, which are enlightening the photographical classification of hailstones prevailed recently in India.

Figure 3.9. Pictures of hailstones prevailed in Baramati, Beed, and Wasim districts of Maharashtra state in India during February-March 2014.

Source: *Photos from collection of Dr. Pralhad Jaybhaye and from multimedia.*

Figure 3.10. Hailstones/beds/sheets on different terrains of Maharashtra state (India) was observed during March 9–16, 2015.

Source: *Photos from collection of Dr. Pralhad Jaybhaye and from multimedia.*

Figure 3.11. Hailstones/beds/sheets on different terrains of Maharashtra state (India) was observed during March 8, 2017.

Source: *Photos from collection of Dr. Pralhad Jaybhaye and from multimedia.*

3.5.4. Snowflakes Mechanism

Tiny supercooled cloud droplets freeze to form the snow crystals. When a droplet is frozen, it is raised into the supersaturated environment. When the content of water droplets is increased compared to ice crystals, these ice crystals are capable to grow and achieve the size of hundreds of micrometers at the cost of the water precipitations present in the clouds. The Wegener-Bergeron-Findeisen method is the name given to this process. The resultant reduction of water droplets origins the water vapors to evaporate, which means that the growth of ice crystals is at the cost of water droplets (Tözeren, 2000; Klesius, 2007). The resulting big crystals are an effective cause of rain, as these fall over the atmosphere because of their collide, and mass and remain unified to form clusters, or large aggregates. Snowflakes are formed as a result of this process and fall to the ground as the kind of ice particle. Guinness World Records enlist the world's prevalent snowflakes at Fort Keogh in January 1887; which possessed the size of 38 cm. The precise particulars of sticking phenomena are yet to be explored (Lawson, 2009).

The ice has a clear structure. Despite this, light scattering through ice crystal surfaces and deficiencies shows that the crystals often seem white because of diffuse light spectrum reflection through the tiny ice particles. The morphology of the snowflake is depending upon the humidity and temperature of atmosphere during its formation. Hardly, snowflakes can be formed in threefold symmetry—triangular snowflakes—at a temperature of around 2°C (Crump, 1994). The utmost general snow particles are asymmetrical, though near ideal snowflakes may be further general in pictures since they are more visually attractive. Because snowflakes form at diverse rates and in different designs as a result of the changing humidity and temperature in the atmosphere via which they fall on their approach to the ground, no two snowflakes are alike. Snow has the METAR code SN, while snow showers have the designation SHSN (Figure 3.12) (Daukantas, 2006; Roach, 2007).

Figure 3.12. Typical snowflake.

Source: https://phys.org/news/2011-12-mystery-snowflakes.html.

3.5.5. Formation of Diamond Dust or Ice Crystals

Diamond dust is also identified as ice crystals or ice needles. It is formed when the temperature is about −40°C (−40°F) and the air has marginally greater moisture content from upward mixing with colder, surface-based air. They comprise simple hexagonal shaped ice crystals. IC is the METAR code to identify diamond dust global weather forecasts every hour is (Libbrecht, 2001; Intrieri and Shupe, 2004).

3.5.6. Deposition of Occult

Occult deposition takes place when the mist or exceedingly saturated air with water vapors passes over the trees, land, and shrubs and interacts with the leaves causing occult deposition (Unsworth and Wilshaw, 1989) and mist is a state of atmospheric obscurity produced by suspended microscopic water droplets or wet hygroscopic particles. The term used for synoptic purpose when visibility is equal to or exceeds 1 km; the corresponding relative humidity (RH) is greater than 95% (Jaybhaye et al., 2013).

3.6. ORIGINS OF PRECIPITATION

3.6.1. Frontal Activity Phenomena

Dynamic precipitation also known as stratiform befalls as a result of slow rise of air in synoptical systems, for instance over cold fronts, and before warm fronts. The same ascent has been witnessed in comma-head rain patterns about mid-latitude cyclones and around hot cyclones outside of the eyewall (Geerts, 2002; Lyttelton et al., 2007). Various kinds of climates are found along an occluded front, perhaps with thunderstorms, but ordinarily, their path is related to the air mass drying. Occluded fronts are typically formed nearby areas with matured reduced pressure. Precipitation might happen on spiritual bodies as well other than Earth. When it becomes cold, Mars partakes rain that almost certainly partakes in the form of ice needles, instead of rain or snow (January, 2015).

3.6.2. Convective Rain

Convective clouds produce convective rain, or showery precipitation, for instance, cumulus or cumulonimbus congestus. Convective rain (Figure 3.13) falls like showers with promptly fluctuating intensity. It falls above a specific region for a comparatively short duration, as these clouds have an inadequate horizontal level. Convective precipitation mostly occurs in the tropics; though, it is proposed that stratiform precipitation too happens. Convective precipitation is common in the mid-latitudes and is commonly related to baroclinic boundaries like warm fronts, squall lines, and cold fronts (Sukovich et al., 2009).

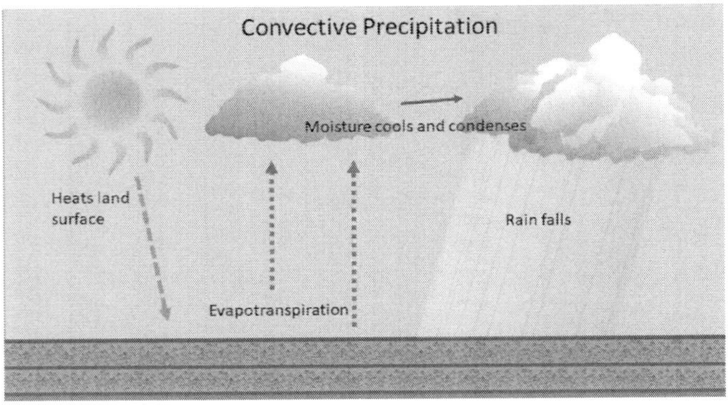

Figure 3.13. The convective rain processes.

Source: *https://massivesci.com/articles/farms-climate-change-crop-rotation-rain-fallow/.*

3.6.3. Orographic Precipitation

Orographic precipitation (Figure 3.14) ensues through the intensifying air movement of a significant moment of moist air on the windward (upwind) side of mountains through the mountain edge, ensuing adiabatic condensation and cooling. Mountainous areas are imperiled to comparatively reliable winds, a high moist climate generally overcomes on the exposed side of a mountain as compared to downwind or leeward adjacent of a mountain. Moisture is eradicated due to orographic lift, and leaves drier air on the descendant and usually warming downwind side at which a rain shadow is perceived (Boutle et al., 2010).

Figure 3.14. The origin of orographic precipitation.

Source: *https://en.wikipedia.org/wiki/Precipitation_types.*

Mount Waialeale, on the Hawaiian island of Kauai, is known for its heavy rainfall, with the 2nd-highest average yearly rainfall on the planet, at 12,000 millimeters. Storm systems influence the state by frequent dense rains from October to March. Limited climates differ significantly on various islands because of their topography as these are isolatable into upwind regions and leeward regions on the basis of the place near higher mountains. Windward sides face intense east to northeast trade winds and acquire considerably more rainfall as compared to leeward sides (*Kona*) which are sunnier and drier, and receive low rain and fewer cloud cover (Koffmann et al., 1996; Goubanova and Li, 2007).

In the Andes Mountain range obstructs appeasing moisture reaching the land, ensues a desert such as climate just downwind through western Argentina. Same effect is generated by Sierra Nevada range generates in North America creating the Mojave Deserts and Great Basin (Knaff et al., 2005). Likewise, the Himalayas Mountains range in Asia generates an obstruction to monsoons and causes elevated rain on the southerly side and low rain on the north side (DeLeon-Rodriguez et al., 2013; Li et al., 2016).

3.6.4. Snow Falling

Extratropical cyclones result in cold and perilous environments with dense rain and snow and high-speed winds (occasionally stated as windstorms in Europe) surpassing 119 km/h (74 mph). The band of precipitation is often widespread related to their warm front, is enforced by the weak upwind above the face boundary when cools off it condenses and produced precipitation in extended band, that is extensive and stratiform, importing altercation out of dark clouds. When mass of arctic air is displaced by moist air a, overrun snow can cause the poleward side of the extended rain band. Poleward is in the North Hemisphere to the North Pole. Nevertheless, poleward is to the South Pole in the Southern Hemisphere (Budyko, 1969; Mares, 2017).

Southwest of extratropical hurricanes, rounded cyclonic movement carrying cold air through the comparatively warm water objects can result in fine lake-effect snow groups. These bands transport strongly localized snowfall that are comprehended as: large water bodies like ponds proficiently store heat causing substantial temperature alterations greater than 13°C among the surface of water and nearby air. As a result of temperature change, hotness, and moisture are transferred upward, which condense into perpendicularly concerned with clouds (as shown in the satellite picture) yielding snow showers (Figure 3.15). Reduction in temperature with altitude and depth of cloud is directly influenced by the large-scale atmosphere and temperature of water. The greater the temperature decline with altitude, the deeper the clouds grow, and the more the rain rate is.

In hilly areas, substantial snowfall gathers as air is enforced to go up the mountains and force out rain near windward hills and in cold environments the precipitation falls as snow. Due to roughness of terrain, predicting the place of dense snowfall becomes an important task (Bosilovich et al., 2002).

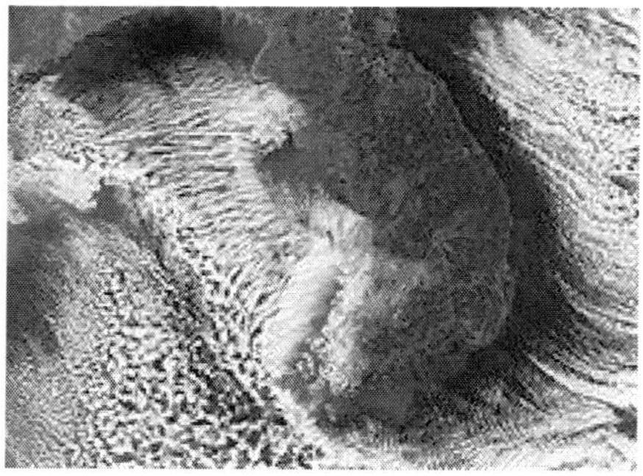

Figure 3.15. The lake-effect snow bands close to the Korean Peninsula in early December 2008.

Source: *https://www.wikiwand.com/en/Precipitation.*

3.7. VARIATIONS IN PRECIPITATION

When the climate fluctuates, the main variations in precipitation are probably in the frequency, intensity, and extent of events, however these physical characteristics are rarely considered in interpretations or models. Why does it rain? When a portion of air upsurges, it enlarges in the lesser pressure, cools, and consequently condenses moisture in the portion of air, creating cloud and, eventually, rainfall—or possibly snowfall. Thus, a significant element involves several and diverse processes for instigating air portion to increase. These extent from orographic elevating, which occurs as air passes over mountains, to a slew of uncertainty in the atmosphere, which rise from uneven heating and potential vorticity dynamics. The uncertainties range from convective instabilities, which directly affect vertical mixing, to meridional heating disparities, which induce baroclinic instabilities, as well as the ubiquitous fronts and low and high-pressure weather systems. As a result, when cold air pushes beneath warmer air (as in a moving cold front) or warm air glides over cold air, and so on, opportunities for air to uplift arise.

Hence, a short term, i.e., "storm," is used for all the prospective disturbances that generate ascending motion in the air, are thunderstorms, or extratropical cyclones expressed as snowstorms and rain. Though the scattering of such events all over earth differs with environment, the elementary mechanisms for triggering air to upsurge do not in and of themselves modify. They require be apprehending and demonstrating better, but from a communal stance, in fact their comparative prominence can be changed with time substantially but sometimes unnoticed circumstances.

Likewise, procedure involving in tangible condensation procedure are of central importance. After a cloud is formed, if the droplets are grown large (> 4-μm diameter) or the air freezes adequately to freeze the cloud droplets, then snow and rain are formed, correspondingly. The microphysics of cloud droplets matter. Human intrusion, like depositing many types of contaminants and aerosols in the atmosphere, can create vital alterations in the size and intensity of within cloud droplets, cloud heating, cloud's lifespan, and precipitation creation. aerosols that absorb Light by deposition of heat could short circuit the hydrological cycle, straight to a layer or the layer is heated through the hidden heating in rain initiating from the absorbed heat at surface then lost via evaporation.

These characteristics are of major concerns and are emphasized in major meteorological and climate study but are not discussed here. The further key element in the primary statement about why it rains is the supposition that there is moisture existing. When it comes to moisture, where does it come from? It is claimed that this characteristic of rain has been less appreciated and is commendable of extra consideration. Despite this, there will be no rain, but if moisture is available, rain will fall. This trait is occasionally overlooked in severe droughts, as the one that occurred in the summer of 2002 in the US, and futile requests for cloud seeding were made in attempt to induce snow and rain. It is unable to generate moisture and does not make the enhanced environment more conducive to the commencement of convection. This may result in greater moisture, but it may come at the expense of precipitation downstream.

Characteristics of rain depend on existing moisture and type of storm. We express persuasively about precipitation, rainfall, and frequently compare it on equality basis with other different atmospheric variables including pressure, wind, temperature, etc. However, it does not always rain. And when it happens, the pace of rain varies. There exist least chances that we may reflect how recurrent and how concentrated the rain is when it happens, along with the entire extents. Continuous, appropriate rains support plants, however the same rainfall amounts in a short time might cause local flooding and run-off, leaving soils significantly drier at the end of the day. The physiognomies of precipitation are equally as significant as the amount, according to this graphic. It may "tune" parameters in models to advance amounts, but only if the amounts are correct for the right explanations—which are a specific combination of two factors, including precipitation amount and frequency. It is unlikely that useful simulations or predictions will emerge.

Furthermore, we face future that the features of rain which is additional pertinent to alteration when weather changes. It shows that expectations are more for variations in the excesses of droughts and floods compared to the all-precipitation extent. These characteristics are imperative for water resources, agriculture, and hydrology however, they are not been sufficiently valued or discussed in the literature of effects brought by climate change. Therefore, a vital tenacity of the current study is to focus the requirement for further consideration to this chapter.

In current chapter, we deliberate problems through the variations to be anticipated in rain, difficulties in modelling it, and want to look at the complications in novel methods; methods that develop understanding and advance models. We emphasize intangible characteristics to elucidate the foremost points, as conclusive data often do not occur.

3.8. MECHANISM OF RAINFALL AND PRECIPITATION

When saturated air is heated, rain falls as a result of a mountain, conventional currents, or frontal action. As it climbs, the saturated air or water vapor cools down. It clings to minute salt, smoke, dust, or seed particles in the atmosphere. Condensation nuclei are the common name for these particles. Condensation happens when cloud droplets adhere to condensation nuclei, resulting in raindrops. As the droplets grow in size, clouds form.

When the clouds grow more, they turn out to be unstable and heavy but undergo cooling at the dry adiabatic rate. This adiabatic rate is defined as for every 1,000 meters rise, the temperature of the water droplets is reduced by 10°C. The precipitation falls to the ground when around 300 meters above the earth's surface, the cloud develops further. The growing clouds develop warmer compared to surrounding air making it unbalanced. As it develops more, it becomes very heavy and falls to the ground as rain. The kind of rain that falls to the ground depends on the concerns responsible for the rising of the saturated air. Each kind of rainfall entails a diverse mechanism that activates the vertical movement of unstable air.

An equipment known as a Rain Gauge (Figure 3.16) calculates the amount of rainfall recorded at a location. A copper/fiber glass reinforced polyesters (FRP) cylinder with a Jar assembly within and a funnel on top serves as the rain gauge (Figure 3.17). To prevent splashing water from entering the gauge, it is set into the ground with only 30 cm of the top above ground level. Rain collects in the jar after falling through the funnel on top of the copper cylinder. After 24 hours, the water is collected and poured into a measuring cylinder where the measurement is taken.

Figure 3.16. Installed FRP rain gauge for measurement of rainfall.

Source: *Jaybhaye et al. (2013).*

Figure 3.17. Different parts of FRP rain gauge and measuring cylinder for measurement of rainfall.

Source: *Jaybhaye et al. (2013).*

3.9. KINDS OF RAINFALL

A considerable ascent of an air mass is necessary to occur condensation and precipitation. There are three main forms of rainfall because its ascent is caused by three different factors. Conventional Rainfall, Orographic or Relief Rainfall, and Frontal or Cyclonic Rainfall are the three main forms. Each of these forms of rainfall has its own characteristics and diagrams that differ from one another as discussed in detail below.

3.9.1. Conventional Rainfall Formation

Rainfall occurs when the air near the earth's surface is heated by the sun. Heating of air makes it lighter. Consequently, the resultant lighter air upsurges cool down and later condenses on the condensation nuclei present in the atmosphere. When an additional rise in water vapor occurs, it converges and moves gradually upwards. This is because there are limited areas to be covered through the converging air. After convergence, it condenses to make thick cumulous clouds. The rising clouds come to be heavier and unstable. This unstable cloud then drops to the ground in the form of raindrops or rainfall (Figure 3.18).

This kind of rainfall often falls in West Africa trailed by lightning and rainstorms as its associated character.

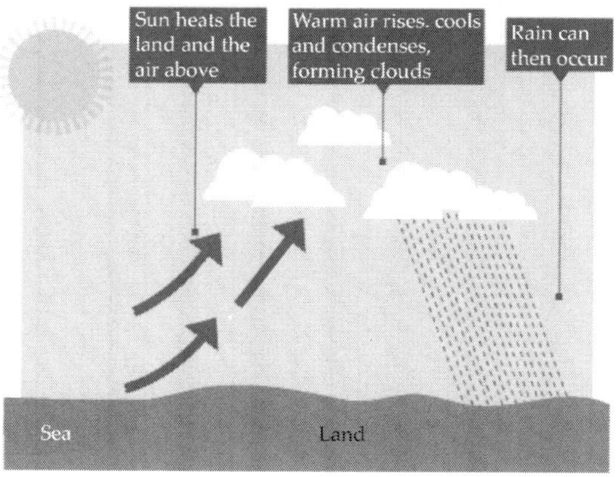

Figure 3.18. The schematic illustration of conventional rainfall formation.

Source: *https://anessoriginal.wordpress.com/elective-geog/variable-weather-and-changing-climate-a-continuing-challenge/key-question-1-wc/1-convectional-rain/.*

3.9.2. Orographic Rainfall

The mountain raises the humid air masses high into the atmosphere when wind pushes moist air to hilly land. The air cools once it has climbed, allowing precipitation to occur (Figure 3.19). They become unstable and heavy when the wind or water vapor rises. They create dense clouds by forming condensational nuclei. They climb much higher, becoming insecure water droplets. As raindrops, they fall to the ground. This type of rainfall is most common on the ground facing the hill or mountain, known as the windward side. The other side, often known as the leeward side, collects descending dry air with little or no rain. It occasionally falls in the form of drops or a shower. Orographic rainfall is most common in hilly places and along hillside slopes. The mountain or hill blocks rain-bearing winds or water vapor and compels it to soar upward. The mountain, in particular, acts as a barrier, forcing the vapor to climb and ultimately resulting in orographic Rainfall.

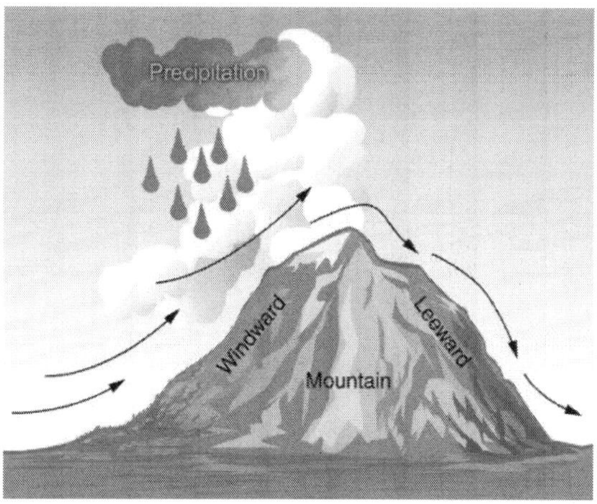

Figure 3.19. The formation of orographic or relief rainfall.

Source: http://www.lcwu.edu.pk/ocd/cfiles/Environmental%20Sciences/Maj/Env-409/Precipitationformsandtypes.pdf.

3.9.3. Cyclonic Rainfall

Lastly, the last form of rainfall is cyclonic rainfall, often known as frontal rainfall. This takes place when two air masses of dissimilar features meet together (Figure 3.20). Such as, when a warm maritime air mass comes across a cold air mass, the cold air mass under-cuts the warm air mass. The cold air mass forces the warm air mass (lighter) to rise leading the warm water vapor to cool down as it rises. The rising air condenses and forms the clouds on the condensation nuclei in the atmosphere. Further rising of the clouds makes them unstable because of the accumulation of more water droplets. As cyclonic rain, they descend to the ground (Shimizu et al., 1993; Semenov and Bengtsson, 2002).

The distinctive characteristics of such rainfall is that it is common in the Tropics, and the Temperate Zone.

Figure 3.20. Displays the creation of cyclonic or frontal rainfall.

Source: *https://www.researchgate.net/figure/showing-the-formation-of-cyclonc-or-frontal-rainfall-Source-Google-search_fig2_322924903.*

3.10. RAINFALL EFFECTS

Water is one of the most precious resources on the planet. Rainwater fills reservoirs, which provide drinking water, give habitat for fish, and provide water to the soil for growth. Harvesting and feeding people and animals are both aided by cultivating crops during the farming season. Rain is regarded to be the most valuable source of water for use in our homes, as well as producing tropical humility for harvesting (Short et al., 1993a, b).

Rainwater, nevertheless, can also have an undesirable result on the Earth when it produces erosion or when it possesses high pH and its typical features which are lightning and thunder. Moreover, it results in a usual disaster which abolishes properties and at times, takes many lives, it causes leaching to such a depth that, plant roots have vanished in the procedure. Factors Intricating in the Characteristics of Precipitation is deliberated below (Soden et al., 2002; Sorooshian et al., 2002):

3.11. SOURCES OF MOISTURE

So, what causes rain? The precipitable water in midlatitudes is characteristically 25 mm and this is also near to the worldwide average value. So how can we probably acquire the quantity greater than an inch of rainfall? The proficiency of mechanisms of rainfall is not that pronounced, possibly 30%, as not all storm-ingested water vapors are transformed to rain and it is not likely to remove moisture from the air Instead, because there are dry downdrafts but also moist cloud debris that stays, the RH of the air residual is typically around 70%. As a result, it is possible that just 7.5 mm of the precipitable water is available for rain. Nonetheless, it can and does rain this much in an hour, and, as we'll see below, the average worldwide rain rate is probably around 45 mm per day, so approximately 6 times the local moisture can be poured in 24 hours (Suppiah and Hennessy, 1996; Trenberth, 1998).

The worldwide average rain rate from global precipitation approximates is around 2.8 mm day^{-1}. Therefore, this is also the worldwide evaporation rate, as the difference is minute. Furthermore, evaporation is constant, dependent on the accessibility of moisture from the surface, and rises to around 5 mm day^{-1} in summer. As a result, the moisture needed for adequate or heavy precipitation does not come via evaporation. As an alternative, it has to originate from the transportation and hence through the union of low-level moisture somewhere else in the atmosphere. These shards of evidence are settled when we realize that it rains only about 5% to 10% of the time on average. Rainfall frequency varies substantially depending on the region and threshold used to detect precipitation; Trenberth (1998) estimates are for 2° latitude by 2.5° longitude grid squares, not point observations. Nondrizzle precipitation values for December-January-February (DJF) and June-July-August (JJA) was used by Dai (2001a) to regulate the rate of different types of precipitation (JJA) are shown in Figure 3.21. At the time of surveillance this approximation joined categories of precipitation with those in the past hour, thus possibly slightly overvaluing the frequency. The zonal means range from around 30% poleward of 60° latitude to roughly 5%–10% in the Tropics. Though these numbers are not capable to constitute a measurable verge, and henceforth it is difficult to determine the global average frequency of occurrence (Trenberth and Guillemot, 1994, 1996, 1998). The area of precipitation, which Burlutskiy (2000) estimates at 5% based on his data, is almost comparable. This, too, is dependent on the threshold and is disproportionately low when compared to Dai (2001a). A sensible working value is possibly ~7% or 1/16 of the globe. It suggests there is a factor of roughly 16 differences between the time-averaged rain rate and the concrete rate based on when it is raining. This fact alone emphasizes the importance of not only the amount of moisture in the atmosphere, but also its location. The previous is influenced by temperature, which controls the atmosphere's moisture-holding size as well as RH (Trenberth, 1999a, b).

We can take the square root of this factor of 16 or so to change it from an aerial degree to a linear degree of system sizes. This means that rainfall-producing weather structures, on average, stretch to distances of 3–5 times the radius of the precipitating zone and collect moisture over that area. This is a fascinating amount and a question is a level to which this does, indeed, diverge from small-scale systems to large-scale systems *on average*; obviously, it does fluctuate extensively for different synoptic environments and individual systems. For a minor thunderstorm with a diameter of 4 km and a radius of 2 km, the moisture originates in a zone around 8 km away and is pushed into the storm by low-level convergence and thunderstorm-scale circulation. Obviously, the region may be uneven in shape due to prevailing winds and shear (Trenberth and Hurrell, 1994). As part of the storm-scale circulation of an extratropical cyclone with rainfall over an 800-kilometer radius, storms across the northern plains of the United States approach and tap moisture from the Gulf of Mexico.

Storms over the northern plains of the United States reach and tap moisture from the Gulf of Mexico as part of the storm-scale circulation of an extratropical cyclone with rainfall over a radius of about 3200 km—or about 30° latitude—and so storms over the northern plains of the United States reach and tap moisture from the Gulf of Mexico as part of the storm-scale circulation of an extratropical cyclone with rainfall over a radius of about. It takes 24 to 48 hours for the moisture to pass through these areas, which is in line with the storm's duration. While hurricanes and some quickly deepening maritime cyclones can greatly enhance the surface flow of moisture coming into the storm, air twists in from huge distances, absorbing moisture on scales far larger than the areas of intense rain.

We estimate that around 70% of the moisture in an extratropical cyclone comes from moisture already present in the atmosphere at the time of the storm's beginning, with the remaining moisture coming from surface evaporation (Trenberth 1998). Because a thunderstorm cell has such a short lifespan, almost all of the moisture is still in the sky when the storm starts (Dabberdt and Schlatter, 1996).

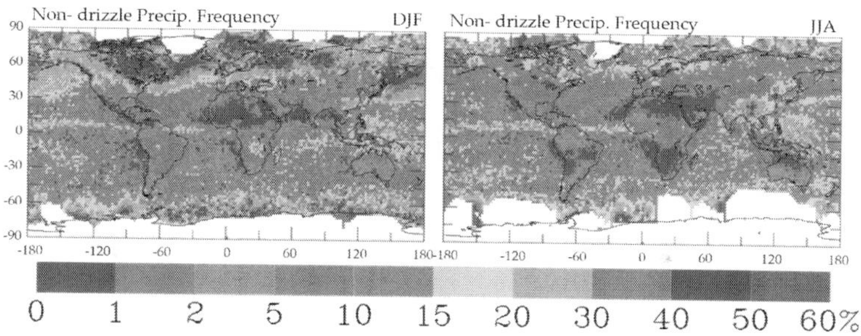

Figure 3.21. The estimated frequency of occurrence (%) of non-drizzle precipitation for (left) DJF and (right) JJA (Dai, 2001a).

Source: *https://agupubs.onlinelibrary.wiley.com/doi/abs/10.1029/1998GL900318.*

The ratio of precipitation amount originating from a limited region via evaporation versus precipitation obtained from advection into the region is called the "recycling" ratio. It differs dramatically after lower winter values to higher summer ones, when large-scale transferences are less important. Observational data, as well as models and isotopic capacities in rain and predicted in replicas based on evaporation in rainfall and fractionation, have all been used to analyze recycling. Each method has disadvantages and advantages related to assumptions, modelling parameterizations, and the capacity to observe isotopic ratios in precipitation. According to Trenberth (1999b), global recycling averages around 10% at 500-km scales, and a map of the expected values for 1000-km scales is shown in Figure 3.22. As potential evapotranspiration (ET) rises in summer, advection, and atmospheric dynamics protrude even more in winter, the yearly average of 6.6% over the Mississippi basin on 500-km distances varies from 3.1% in DJF to 9.3% in JJA. Trenberth (1999b) calculated a 21% annual recycle rate for the basin. Despite the fact that data vary depending on the approach employed, dry conditions in late spring are clearly favorable to drought increasing, enduring, and becoming maintained over the summer due to the need for moisture from local evaporation. In the summer of 2002, there was a severe drought over the United States' intermountain region (Burlutskiy, 2000; Carbone et al., 2002).

This study exhibits the necessity to precisely measure and tracks moisture accessibility in the atmosphere. The Gulf of Mexico, the subtropical North Pacific, and the subtropical and tropical Atlantic are the source regions for the United States (Trenberth 1998). As the storm-scale circulation collects the moisture, tracking moisture migration helps to reveal the large-scale processes at action. Hence, it is probable to trail the large-scale low-level union and deviation designs in the atmosphere.

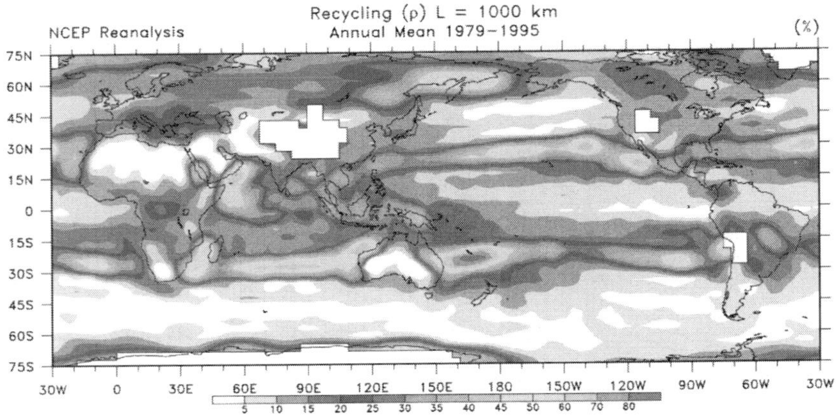

Figure 3.22. Within a 1000 km length scale, an estimation of the annual mean recycling ratio of the percentage precipitation that comes from evaporation (Trenberth, 1999b).

Source: *https://link.springer.com/article/10.1023/A:1005319109110.*

3.12. DIURNAL CYCLE IN PRECIPITATION

A method to discourse the concerns regarding intensity and frequency of rain is to methodically inspect the timing and interval of rain actions as a function of time of day. This permits the systematic errors in forecasting start time and duration of rain in models to be exposed. Therefore, the day cycle lets us instigate to come to grasps with the features of precipitation and how well they are shown (Trenberth and Hoar, 1996). The diurnal cycle in rain is principally prominent over the United States in summer (Figure 3.23) and is poorly simulated in most arithmetical models. The average pattern of the diurnal cycle of summer U.S. precipitation is considered by late afternoon maxima over the Rocky Mountains and the Southeast, and midnight maxima over the region east of the Rockies and the neighboring plains (Emanuel and Coauthors, 1995; Eltahir and Bras, 1996).

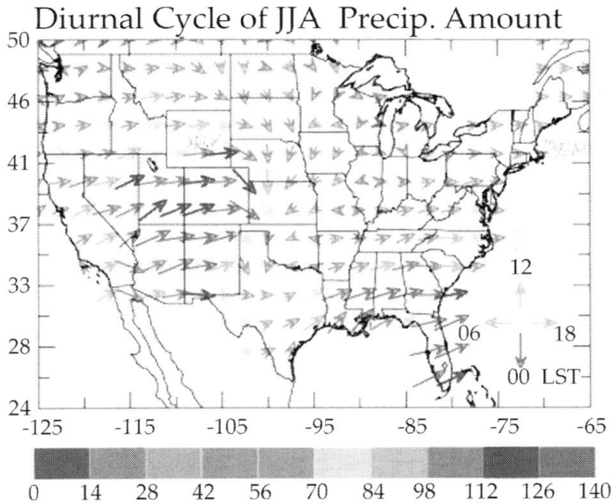

Figure 3.23. The diurnal cycle of total precipitation amount using hourly data for JJA for the United States.

Note: The direction denotes the timing of maximum in local standard time, as indicated by the key in the lower right, and the vector length and color indicate the amplitude as a percent of the diurnal mean. The amount of the diurnal cycle is nearly entirely determined by changes in frequency (timing) rather than intensity.

Source: *https://agupubs.onlinelibrary.wiley.com/doi/abs/10.1029/1998GL900318.*

In other seasons, diurnal fluctuations in precipitation are weaker, in the winter, most of the United States experiences early to late morning maxima. Precipitation frequency has a diurnal cycle is responsible for the majority of diurnal fluctuations, although diurnal variations in precipitation intensity are minor. Significant largescale convergence is associated to the solar-driven diurnal and semidiurnal cycles in surface pressure over a maximum of the western and eastern US during the evening and afternoon, and above the region east of the Rockies at night (Trier and Parsons, 1993); Trenberth and Caron, 2000). Dai et al. (1999) found that the diurnal cycle of low-level large-scale convergence is consistent with daytime convection suppression and a preference for nighttime moist convection over the region east of the Rockies and related plains. Over the central plains, diurnal differences in climatologically significant circulations, such as the low-level jet and north-south baroclinic gradients, interact to provide a nighttime environment with little inhibition to convective start and high levels of convective instability. The development of the low-level jet off the Gulf of Mexico, as well as the eastward extension of late afternoon thunderstorms formed over the Rockies, improve the nighttime maxima in the region east of the Rockies. In the summer, afternoon moist convection is favored by both static instability and surface convergence across the southeast and the Rockies, resulting in large late-afternoon precipitation maxima (Wallace et al., 1996; Vuille et al., 2003). The distinguishing but complex diurnal atmospheric moisture cycle over North America has been documented by Dai et al. (2002) by utilizing high-temporal-resolution Global Positioning System data, and current certification of the progression of the daytime precipitation cycle over the US from a synoptic perspective is specified by Carbone et al. (2002). Models can characteristically pretend some but not all of the daytime cycle design, but certain models are incorrect universally. For example, the operational model of the European Center for Medium-Range Weather Forecasts inclines to produce maximum rain around local noon, which corresponds to maximum heating, and this is also true in several other models. This is about 3–4 hours earlier than what occurs in nature. The diurnal precipitation occurs roughly 2 hours before it does in nature in the NCAR Community Climate System Model and the related atmospheric module, and the complicated structure of nocturnal maxima in the United States Great Plains is missing (see Figure 3.24). Figure 3.24 also demonstrates the distinguishing nature of the diurnal cycle in several other parts of the globe. Over numerous portions of the oceans, there is frequently a minimum in precipitation from late afternoon to about midnight, which is out of phase with maximum continental areas.

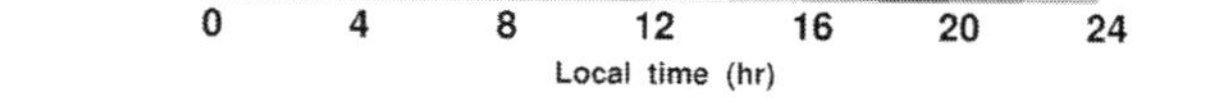

Figure 3.24. The local solar timing (h) of the maximum of the diurnal cycle of precipitation for JJA for convective precipitation is shown in (left) data for 1976–97 (Dai 2001b) vs. (right) a control run (10-year avg) of the CCSM.

Source: *https://journals.ametsoc.org/view/journals/clim/14/6/1520-0442_2001_014_1112_gpatfp_2.0.co_2.xml.*

Dai et al. (1999) showed that model convection techniques created very much mistiness over the Southeast, reducing surface solar radiation and changing the midday peak heat at the surface. Because the model's starting conditions for humid convection are too lax, moist convection occurs too early and too frequently in the model. Early triggering promotes weaker convection and precipitation, as well as weaker downdrafts and gust fronts, in the real world. Because of the prominent role gust fronts play in inciting convection, subsequent convection is expected to be weaker. Premature convection and cloudiness impair adequate heating at the continent's surface, preventing the formation of a continental-scale "sea breeze" and its related convergence and divergence patterns. Moisture migration and its role in convective instabilities are altered as a result. Figure 3.25 depicts the distinct processes, as well as the relevant model errors and responses (Wang et al., 2002; Wilby and Wigley, 2002).

3.13. ATMOSPHERIC CIRCULATION VARIATIONS

For decadal and interannual changeability of precipitation, an additional imperative feature is the part of atmospheric movement alterations. Variations in natural modes of atmospheric motion have been recognized and may be associated with anthropogenic climate alteration (Figure 3.25). The Pacific-North American (PNA) teleconnection pattern, the North Atlantic Oscillation (NAO), and the El Nio-Southern Oscillation (ENSO) all interact to affect the planetary wave structure over the Northern Hemisphere, causing extreme wintertime temperatures to rise over Eurasia and North America rather than cool over the northern oceans in recent years.

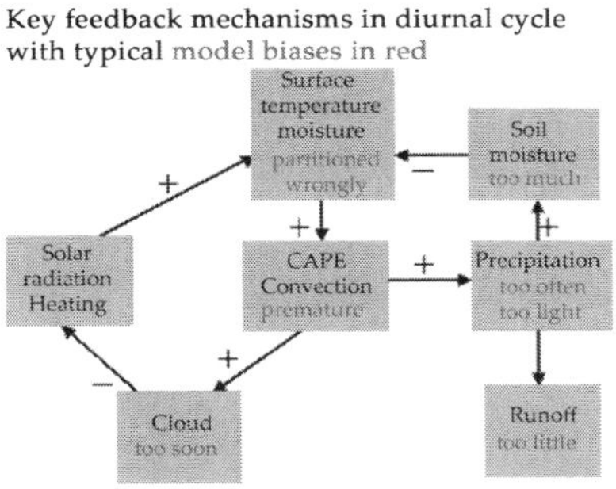

Figure 3.25. The mechanisms involved in the diurnal cycle along with the model biases (in red).

Source: *https://www.researchgate.net/publication/235640653_The_Changing_Character_of_Precipitation.*

For most of the 1980s and 1990s, the NAO index was abnormally high (Hurrell 1995), and such trends appear to be linked to tropical Pacific and Indian Ocean warming (Hoerling et al., 2001). During the same time period, ENSO has showed a statistically significant bias favoring the El Nio phase. A shift in storm paths connected to teleconnections is one way that rainfall patterns in the midlatitudes can change. Because the NAO has been in a more upbeat mood, a dipole pattern of change has emerged over Europe, with lower rainfall in southern Europe and wetter conditions in Scandinavia (Hurrell, 1995; Wilson and Schreiber, 1986). In the Pacific, distinct changes take place in storm tracks over the North Pacific in relationship with ENSO and the PNA resulting in a dipole pattern of precipitation irregularities that encompasses California has a factor over the southeastern United States at times. Therefore, there is higher storm track action and rainfalls to the south and reduced rainfalls to the north. Furthermore, droughts, and floods in various locations around the globe are related to ENSO through teleconnections. Guillemot and Trenberth (1996) show how storm paths shifted over North America during the 1988 spring-summer drought and the 1993 floods. ENSO causes dipole "boomerang" shaped structures in the tropical Pacific as the Inter-Tropical Convergence Zone and South Pacific Convergence Zone move together. The predictions for fluctuations in atmospheric flow, whether associated with alterations in sea surface temperatures or not, are actual and add significant difficulty to prospective changes in rain amount locally. Nonetheless, variations in features of rain could be further vigorous (Xie and Arkin, 1997; Wright et al., 2001).

3.14. TRENDS IN PRECIPITABLE WATER

After 1973, the amount of moisture in the atmosphere is usually detected to be increasing. Annual mean precipitable water amounts below 500 mb improved by about 5%in a decade as a statistically significant trend in the Western Hemisphere north of the equator from 1973 to 1995 (Ross and Elliott 1996), and these relate to momentous escalations of 2%–3% in a decade in RH over the Southeast, Caribbean, and subtly influenced Hawaii (Ross and Elliott 1996). The majority of the rise is due to a rise in temperature and, as a result, an increase in atmospheric water-holding capacity. Zhai and Eskridge (1997) discovered increased trends in precipitable water in all seasons and for the annual mean in China from 1970 to 1990. Increases in moisture over the western Pacific had already been documented by Hense et al. (1988). Precipitable water and RH have not increased throughout much of

Canada, and there have been noticeable reductions where temperatures have dropped in northeast Canada. To conclude, while reservations occur because of inaccuracies in humidity measurements, maximum studies specify that water vapor upsurges are existing in numerous regions.

We previously discussed how atmospheric moisture levels over adjacent oceans affect precipitation in the United States, and precipitable water levels in all of these places have risen by more than 10% since 1973. When it rains, this should result in a 10% increase in rainfall rates if all other conditions are equal, because low-level moisture convergence will be improved by that amount. Increased atmospheric moisture content is thought to favor greater rainfall and snowfall events, increasing the risk of flooding. In the United States there is a vibrant indication that rainfall rates have altered, After 1910, for example, in an area with total annual precipitation of more than 2-in (50.8 mm) quantities from 1-day extremes. Over the 20th century, the "much above normal" area, defined as the upper 10% inclusive, improved progressively from less than 9% to over 11%, a 20% relative change in total. Further analysis of precipitation rises in the United States by Karl and Knight (1998) revealed that it occurs frequently in the upper 10th percentile of the distribution and that the proportion of total precipitation resulting from extreme and heavy events increased at the expense of more moderate events. According to Kunkel et al. (1999), extreme precipitation occurrences lasting 1–7 days improved at a rate of roughly 3% per decade in the United States from 1931 to 1996. Trenberth (1998) and Dai (1999) presented patterns of change in the frequency of hourly precipitation across the United States, as well as El Nio correlations. Japan (Iwashima and Yamamoto 1993) and Australia provide additional evidence for rising precipitation rates. While higher rainfall rates increase the likelihood of flooding, local governments, the Corps of Engineers, and the Bureau of Reclamation in the United States routinely ameliorate floods, and flood records are frequently confounded by changes in land use and increased human habitation in flood plains. However, excessive floods are growing in the 20th century. Nevertheless, the foremost way to assess these concepts would be to track the hourly rain data (Hennessy et al., 1997).

3.15. PRECIPITATION CHANGE AS THE CLIMATE CHANGES

At the surface, in the atmosphere variations in radiative compelling linked with growing greenhouse gases create amplified heating. We undertake climate variations related to aerosols are allocated with somewhere else. The total of heating is determined analytically on all kinds of responses, together with ice-albedo feedback, effects of changes in clouds, and water vapor feedback. Uncertainties arise as a result of climate models' insensitivity to a doubling of CO_2 un global mean surface temperature of 1.5°C to 4.5°C. Though, according to the Intergovernmental Panel on Climate Change (IPCC) models, the overall change in the hydrological cycle (e.g., total precipitation) will be around 1%–2% K1. The quality of this range has not been determined, and specified concerns about how effectively models imitate the diurnal cycle, it could be extremely misleading.

More surface evaporation would be expected to decrease temperature rises since evaporation at the surface cools the globe and so acts to "air condition" it. Indeed, with global warming, an increase in potential ET is a very robust finding in all climate models (IPCC 2001). When there is no precipitation, the risk of drought increases because surface drying improves. In conjunction with such droughts, it also increases the risk of heatwaves and wildfires, because once the soil moisture is depleted, all of the heat is directed toward rising temperatures and wilting plants (Zhai and Eskridge, 1997).

Nevertheless, a very well-defined number is the addition of around 7% K1 to the atmosphere's water-holding capacity, as controlled by the Clausius-Clapeyron equation. Furthermore, models suggest that RH variations are minimal, owing to precipitation physics, and this is supported by limited data. As a result, the concrete moisture content of the atmosphere should likewise rise at a similar rate, which is consistent with interpretations in areas of the world where significant pollution clouds are absent. Nevertheless, it is improbable that the moisture fluctuations will be unvarying even though models envisage escalations in surface temperatures almost everywhere after a few decades with proliferation in greenhouse gases. Atmospheric dynamics perform a role through preferred regions of convergence and subsidence. Despite the fact that higher latitudes are expected to see larger increases in surface temperatures, Clausius-nonlinear Clapeyron's relationship with temperature suggests that lower latitudes will see a larger absolute increase in moisture amount. As a result, much of the moisture may be beyond the reach of extratropical storms. As a result, examining the generality of this heuristic reasoning is a research subject. Nonetheless, tracking the consequences provides a conceptual foundation for inferring observations and models.

Because heavy rainfall rates far surpass evaporation rates and thus rely on low-level moisture convergence, we hypothesized that rainfall intensity should increase at roughly the same rate as moisture, namely 7% K1 with warming. The additional increasing convergence of moisture may even exceed this rate of rise since the more latent heat released feeds back and revitalizes the storm that created the rain in the first place. This means that when it rains, the rain rate variability contradicts the 1–2% K1 for total rainfall amounts. According to Hennessey et al. there must be a reduction in moderate and light rainfall, as well as a decrease in the frequency of rain occurrences (1997). As a result, rainfall—or snowfall—events may be less frequent but more intense.

3.16. APPREHENSIONS AND PROBLEMS

The change of climate is very probable to locally alter the frequency, amounts of precipitation, duration, and intensity. Analysis of how successfully climate models handle certain precipitation characteristics is a concern of substantial social importance. The primary necessity is improved documentation and handling of all characteristics of precipitation. Using hourly precipitation amounts, for the creation of a database of intensity and frequency Trenberth (1998) has debated. It works well with global models' time increments, which are typically two or three each hour. The average time of convective episodes in the Tropics, according to Ricciardulli and Sardeshmuhk (2002), is 5.5 hours. This time interval is now averaged over individual cells in a storm, yet it allows for a complete understanding of the storm's evolution. As per a data management stance, it is possible. It can also be deduced from numerous explanations, such as rain gauge recordings and radar. Mohr et al. (1999) and Zipser and Mohr (1996) utilized Special Sensor Microwave Imager 85-GHz data to designate intensity, size, and geographic distribution of mesoscale convective systems and clusters of cloud and their influences on rainfall. Nesbitt et al. (2000) employed diverse sensors on the TRMM to recognize and categorize advancing clouds and to deliver an understanding of the intensities of rain and lightning (National Research Council (NRC), 1998; Nesbitt et al., 2000).

Supplementary issues comprise an enhanced knowledge of the proficiency of rain and how it varies with environmental circumstances. Vertical shear, entrainment, water loading, instability, and system type all influence efficiency, and aerosol and microphysical impacts on rain, as explained in the 1st section, may also play a role. Such queries necessitate a thorough understanding of water vapor transfer in storms. Figure 3.26 is an illustration of the incomprehensible richness of water vapor's changeability in the atmosphere (Numaguti, 1999).

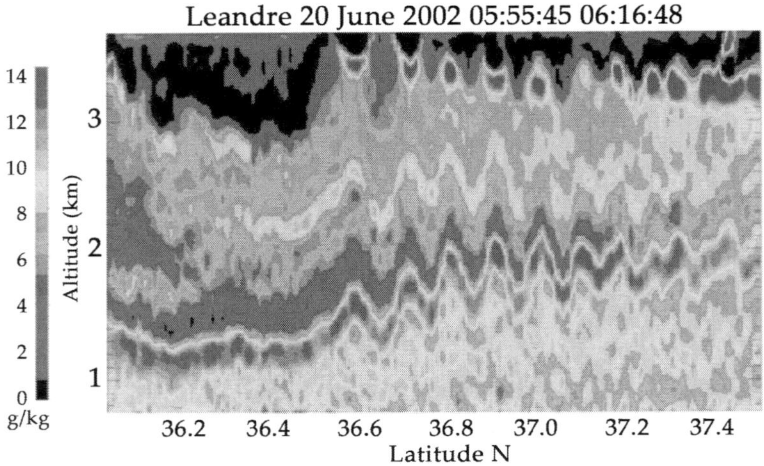

Figure 3.26. Water vapor variability from the IHOP 2002 project.

Note: The picture depicts a vertical cross-section of the water vapor mixing ratio as determined by the French LEANDRE 2 water vapor DIAL (differential absorption lidar) aboard the P-3 aircraft of the US Naval Research Laboratory. As the aeroplane proceeded near a nocturnal convective system, the cross-section is oriented N-S. In reaction to the system's circulations, the moisture deepened with wavelike perturbations.

Source: *https://www.researchgate.net/publication/235640653_The_Changing_Character_of_Precipitation.*

REFERENCES

1. Abbott, B. W., Bishop, K., Zarnetske, J. P., Minaudo, C., Chapin, F. S., Krause, S., & Pinay, G., (2019). Human domination of the global water cycle absent from depictions and perceptions. *Nature Geoscience, 12*(7), 533–540.
2. Anagnostou, E. N., (2004). A convective/stratiform precipitation classification algorithm for volume scanning weather radar observations. *Meteorological Applications: A Journal of Forecasting, Practical Applications, Training Techniques and Modelling, 11*(4), 291–300.
3. Anonymous, (2014). *Daily Agrowon, 1, 1-10.*
4. Biswas, B. S., & Marar, A., (2018). *Reported, 1.24 Lakh ha. in 11 Districts Affected*. Available on: https://www.thehindian express.com (accessed on 18 August 2021).
5. Bosilovich, M. G., & Schubert, S. D., (2002). Water vapor tracers as diagnostics of the regional hydrologic cycle. *J. Hydrometeor., 3*, 149–165.
6. Boutle, I. A., Beare, R. J., Belcher, S. E., Brown, A. R., & Plant, R. S., (2010). The moist boundary layer under a mid-latitude weather system. *Boundary-Layer Meteorology, 134*(3), 367–386.
7. Braham, Jr. R. R., (1964). What is the role of ice in summer rain-showers? *Journal of Atmospheric Sciences, 21*(6), 640–645.
8. Braun, L., (2010). Limitations of CAPE index for use in central Europe. *Meteorologický Časopis*, 127–132.
9. Brimelow, J. C., Reuter, G. W., & Poolman, E. R., (2002). Modeling maximum hail size in Alberta thunderstorms. *Weather and Forecasting, 17*(5), 1048–1062.
10. Brubaker, K. L., Entekahabi, D., & Eagleson, P. S., (1993). Estimation of precipitation recycling. *J. Climate, 6*, 1077–1089.
11. Budyko, M. I., (1969). The effect of solar radiation variations on the climate of the *Earth. Tellus, 21*(5), 611–619.
12. Burlutskiy, R. F., (2000). Diagnosis of the energy cycle of the Northern Hemisphere from the analysis of Hydrometcenter of Russia and ECMWF reanalysis. *Proc. Second WCRP Int. Conf. on Reanalysis* (Vol. 985, pp. 390–393). WCRP109, WMO/TD-No. 985, Wokefield Park, United Kingdom, WCRP.

13. Businger, S., & Reed, R. J., (1989). Cyclogenesis in cold air masses. *Weather and Forecasting*, *4*(2), 133–156.
14. Carbone, R. E., Tuttle, J. D., Ahijevych, D. A., & Trier, S. B., (2002). Inferences of predictability associated with warm season precipitation episodes. *J. Atmos. Sci., 59*, 2033–2056.
15. Cavallo, S. M., & Hakim, G. J., (2009). Potential vorticity diagnosis of a tropopause polar cyclone. *Monthly Weather Review*, *137*(4), 1358–1371.
16. Conway, G. A., Hill, A., Martin, S., Mode, N. A., Berman, M. D., Bensyl, D. M., & Moran, K. A., (2004). Alaska air carrier operator and pilot safety practices and attitudes: A statewide survey. *Aviation, Space, and Environmental Medicine*, *75*(11), 984–991.
17. Crump, R., (1994). Snowflake science. *The Science Teacher*, *61*(1), 45.
18. Dabberdt, W. F., & Schlatter, T. W., (1996). Research opportunities from emerging atmospheric observing and modeling capabilities. *Bull. Amer. Meteor. Soc., 77*, 305–324.
19. Dai, & Deser, C., (1999). Diurnal and semidiurnal variations in global surface wind and divergence fields. *J. Geophys. Res., 104*, 31 109–31 125.
20. Dai, (2001a). Global precipitation and thunderstorm frequencies. Part I: Seasonal and interannual variations. *J. Climate, 14*, 1092–1111.
21. Dai, (2001b). Global precipitation and thunderstorm frequencies. Part II: Diurnal variations. *J. Climate, 14*, 1112–1128.
22. Dai, A., (1999). Recent changes in the diurnal cycle of precipitation over the United States. *Geophys. Res. Lett., 26*, 341–344.
23. Dai, Giorgi, F., & Trenberth, K. E., (1999). Observed and model simulated precipitation diurnal cycles over the contiguous United States. *J. Geophys. Res., 104*, 6377– 6402.
24. Dai, Wang, J., Ware, R. H., & van Hove, T., (2002). Diurnal variation in water vapor over North America and its implications for sampling errors in radiosonde humidity. *J. Geophys. Res., 107*, 4090, doi:10.1029/2001JD000642.
25. Daukantas, P., (2006). Physicist's snowflakes grace holiday stamps. *Optics and Photonics News*, *17*(12), 8–11.
26. DeLeon-Rodriguez, N., Lathem, T. L., Rodriguez-R, L. M., Barazesh, J. M., Anderson, B. E., Beyersdorf, A. J., & Konstantinidis, K. T., (2013). Microbiome of the upper troposphere: Species composition and

prevalence, effects of tropical storms, and atmospheric implications. *Proceedings of the National Academy of Sciences, 110*(7), 2575–2580.

27. Dhingra, R. D., Jennings, D. E., Barnes, J. W., & Cottini, V., (2021). Lower surface temperature at bright ephemeral feature site on Titan's north pole. *Geophysical Research Letters*, *48*(7), e2020GL091708.
28. Dore, A. J., Mousavi-Baygi, M., Smith, R. I., Hall, J., Fowler, D., & Choularton, T. W., (2006). A model of annual orographic precipitation and acid deposition and its application to Snowdonia. *Atmospheric Environment*, *40*(18), 3316–3326.
29. Eltahir, E. A. B., & Bras, R. L., (1996). Precipitation recycling. *Rev. Geophys., 34*, 367–378.
30. Emanuel, K., & Coauthors, (1995). Report of the first prospectus development team of the U.S. Weather Research Program to NOAA and the NSF. *Bull. Amer. Meteor. Soc., 76*, 1194–1208.
31. Ferguson, C. R., & Wood, E. F., (2011). Observed land-atmosphere coupling from satellite remote sensing and reanalysis. *Journal of Hydrometeorology*, *12*(6), 1221–1254.
32. Glickman, T. S., & Zenk, W., (2000). *Glossary of Meteorology* (Vol. 1, pp. 1–23). AMS (American Meteorological Society).
33. Goubanova, K., & Li, L., (2007). Extremes in temperature and precipitation around the Mediterranean basin in an ensemble of future climate scenario simulations. *Global and Planetary Change, 57*(1–2), 27–42.
34. Graves, S. D. B., McKay, C. P., Griffith, C. A., Ferri, F., & Fulchignoni, M., (2008). Rain and hail can reach the surface of Titan. *Planetary and Space Science*, *56*(3–4), 346–357.
35. Hennessy, K. J., Gregory, J. M., & Mitchell, J. F. B., (1997). Changes in daily precipitation under enhanced greenhouse conditions. *Climate Dyn., 13*, 667–680.
36. Hense, A., Krahe, P., & Flohn, H., (1988). Recent fluctuations of tropospheric temperature and water vapor content in the tropics. *Meteor. Atmos. Phys., 38*, 215– 227.
37. Houze, Jr. R. A., (1997). Stratiform precipitation in regions of convection: A meteorological paradox? *Bulletin of the American Meteorological Society*, *78*(10), 2179–2196.
38. Hurrell, (1996). Influence of variations in extratropical wintertime teleconnections on Northern Hemisphere temperature. *Geophys. Res. Lett., 23*, 665–668.

39. Intrieri, J. M., & Shupe, M. D., (2004). Characteristics and radiative effects of diamond dust over the western Arctic Ocean region. *Journal of Climate, 17*(15), 2953–2960.
40. IPCC (2001). In: Houghton, et al., (eds.), *Climate Change: The Scientific Basis* (Vol. 1, p. 881). Cambridge University Press, Cambridge, United Kingdom.
41. Iwashima, T., & Yamamoto, R., (1993). A statistical analysis of the extreme events: Long-term trend of heavy daily precipitation. *J. Meteor. Soc. Japan, 71*, 637–640.
42. January, P., (2015). Stargazer on the rise to becoming an astrophysicist: Unanswered question on how far light travels into space inspires greater interest in astronomy [WIE from around the world]. *IEEE Women in Engineering Magazine*, *9*(2), 56–58.
43. Jaybhaye, P., Raundal, P., & Andhale, P., (2013). *Handbook of Agrometeorology* (pp. 12, 169). Contenental Prakashan, Vijaya Nagar, Pune – 411 030, Maharashtra, India.
44. Jewell, R., & Brimelow, J., (2009). Evaluation of Alberta hail growth model using severe hail proximity soundings from the United States. *Weather and Forecasting, 24*(6), 1592–1609.
45. Johnson, B., & Shepherd, J. M., (2018). An urban-based climatology of winter precipitation in the northeast United States. *Urban Climate, 24*, 205–220.
46. Junker, N. W., (2008). *An Ingredients-Based Methodology for Forecasting Precipitation Associated with MCS's* (Vol. 1, pp. 1–23).
47. Karger, D. N., Conrad, O., Böhner, J., Kawohl, T., Kreft, H., Soria-Auza, R. W., & Kessler, M., (2017). Climatologies at high resolution for the earth's land surface areas. *Scientific Data, 4*(1), 1–20.
48. Karl, T. R., & Knight, R. W., (1998). Secular trends of precipitation amount, frequency, and intensity in the United States. *Bull. Amer. Meteor. Soc., 79*, 231– 242.
49. Karl, T. R., Easterling, D. R., & Quayle, R. G., (1996). Indices of climate change for the United States. *Bull. Amer. Meteor. Soc., 77*, 279–292.
50. Klesius, M., (2007). The mystery of snowflakes. *National Geographic, 211*(1), 20.
51. Knaff, J. A., Sampson, C. R., & DeMaria, M., (2005). An operational statistical typhoon intensity prediction scheme for the western North Pacific. *Weather and Forecasting, 20*(4), 688–699.

52. Koffmann, W., Wade, N. L., & Nicol, H., (1996). Tree sprays and root pruning fail to control rain induced cracking of sweet cherries. *Plant Protection Quarterly*, *11*, 126–130.

53. Kunkel, K. E., Andsager, K., & Easterling, D. R., (1999). Long-term trends in extreme precipitation events over the conterminous United States and Canada. *J. Climate, 12*, 2515–2527.

54. Laws, J. O., (1941). Measurements of the fall-velocity of water-drops and raindrops. *Eos, Transactions American Geophysical Union*, *22*(3), 709–721.

55. Lawson, J. E., (2009). *Hands-on Science: Light, Physical Science (Matter)-Chapter 5: The Colors of Light* (Vol. 1, pp. 1–23). Portage & Main Press. ISBN: 9781894110631.

56. Lee, T. F., Turk, F. J., & Richardson, K., (1997). Stratus and fog products using GOES-8-9 3.9-μ m data. *Weather and Forecasting*, *12*(3), 664–677.

57. Li, N., Cheung, K. F., Stopa, J. E., Hsiao, F., Chen, Y. L., Vega, L., & Cross, P., (2016). Thirty-four years of Hawaii wave hindcast from downscaling of climate forecast system reanalysis. *Ocean Modelling, 100*, 78–95.

58. Liang, X. Z., Kunkel, K. E., & Samel, A. N., (2001). Development of a regional climate model for U.S. Midwest applications. Part I: Sensitivity to buffer zone treatment. *J. Climate, 14*, 4363–4378.

59. Libbrecht, K. G., (2001). Morphogenesis on ice: The physics of snow crystals. *Engineering and Science*, *64*(1), 10–19.

60. Lozowski, E. P., Brett, M., Tait, N., & Smy, T., (1991). Simulating giant hailstone structure with a ballistic aggregation model. *Quarterly Journal of the Royal Meteorological Society*, *117*(498), 427–431.

61. Manor-Chapman, E., (2013). Cassini solstice mission overview and science results. In: *2013 IEEE Aerospace Conference* (Vol. 1, pp. 1–13). IEEE.

62. Mapes, B. E., & Houze, Jr. R. A., (1995). Diabatic divergence profiles in western Pacific mesoscale convective systems. *Journal of Atmospheric Sciences*, *52*(10), 1807–1828.

63. Mares, M. A., (2017). *Encyclopedia of Deserts* (Vol. 1, pp. 1–24). University of Oklahoma Press.

64. Menon, P. A., (1993). *Ways of the Weather* (p. 44). National Book Trust, India.

65. Menon, S., Hansen, J., Nazarenko, L., & Luo, Y., (2002). Climate effects of black carbon aerosols in China and India. *Science, 297*, 2250–2253.
66. Mesinger, F., & Mesinger, N., (1992). Has hail suppression in eastern Yugoslavia led to a reduction in the frequency of hail? *Journal of Applied Meteorology and Climatology*, *31*(1), 104–111.
67. Mohr, Famiglietti, J. S., & Zipser, E. J., (1999). The contribution to tropical rainfall with respect to convective system type, size and intensity estimated from the 85 GHz ice-scattering signature. *J. Appl. Meteor., 38*, 596–606.
68. Mohr, K. I., & Zipser, E. J., (1996). Mesoscale convective systems defined by their 85-GHz ice scattering signature: Size and intensity comparison over tropical oceans and continents. *Mon. Wea. Rev., 124*, 2417–2437.
69. Mossop, S. C., & Carte, A. E., (1957). Cloud physics research in South Africa. *South African Journal of Science*, *53*(11), 293–297.
70. National Research Council (NRC), (1998). *The Atmospheric Sciences: Entering the 21st Century* (Vol. 1, p. 384). National Academy Press.
71. Nesbitt, S. W., Zipser, E. J., & Cecil, D. J., (2000). A census of precipitation features in the tropics using TRMM: Radar, ice scattering, and lightning observations. *J. Climate, 13*, 4087–4106.
72. Numaguti, A., (1999). Origin and recycling processes of precipitating water over the Eurasian continent: Experiments using an atmospheric general circulation model. *J. Geophys. Res., 104*, 1957–1972.
73. Oguntoyinbo, J. S., & Akintola, F. O., (1983). Rainstorm characteristics affecting water availability for agriculture. *Hydrology of Humid Tropical Regions IAHS Publication, 140*, 1–33.
74. Patel, A. A., & Cox, D. N., (2017). Behavioral and functional assays for investigating mechanisms of noxious cold detection and multimodal sensory processing in Drosophila larvae. *Bio-Protocol*, *7*(13), 1–33.
75. Pearce, R. P., (2002). *Meteorology at the Millennium* (Vol. 83, pp. 1–23). Academic Press.
76. Pidwirny, M., (2008). Chapter 8: Introduction to the hydrosphere (e). Cloud formation processes. *Physical Geography* (Vol. 1, pp. 1–23).
77. Ralph, F. M., Dettinger, M. D., Cairns, M. M., Galarneau, T. J., & Eylander, J., (2018). Defining "atmospheric river": How the glossary of meteorology helped resolve a debate. *Bulletin of the American Meteorological Society*, *99*(4), 837–839.

78. Ramanathan, V., Crutzen, P. J., Kiehl, J. T., & Rosenfeld, D., (2001). Aerosols, climate, and the hydrological cycle. *Science, 294*, 2119–2124.

79. Rao, V. U. M., Rao, B. B., Sikka, A. K., Subba, R. A. V. M., Singh, R., & Maheswari, M., (2014). *Technical Bulletin-Hailstorm Threat to Indian Agriculture: A Historical Perspective and Future Strategies.*

80. Rauber, R. M., Olthoff, L. S., Ramamurthy, M. K., & Kunkel, K. E., (2001). Further investigation of a physically based, nondimensional parameter for discriminating between locations of freezing rain and ice pellets. *Weather and Forecasting, 16*(1), 185–191.

81. Ricciardulli, L., & Sardeshmuhk, P. D., (2002). Local time- and space scales of organized tropical deep convection. *J. Climate, 15*, 2775–2790.

82. Roach, J., (2007). 'No two snowflakes the same' likely true, research reveals. *National Geographic News* (Vol. 1, pp. 1–22).

83. Ross, R. J., & Elliott, W. P., (1996). Tropospheric water vapor climatology and trends over North America: 1973–1993. *J. Climate, 9*, 3561–3574.

84. Ross, R. J., & Elliott, W. P., (2001). Radiosonde-based Northern Hemisphere tropospheric water vapor trends. *J. Climate, 14*, 602–1611.

85. Rutledge, G. K., Alpert, J., & Ebisuzaki, W., (2006). NOMADS: A climate and weather model archive at the National Oceanic and Atmospheric Administration. *Bulletin of the American Meteorological Society, 87*(3), 327–342.

86. SANDRP, (2017). *Maharashtra Farmers Face Impacts of hailstorms and State's "Inaction" Plan on Climate Change*| South Asia Network on Dams, Rivers and People (SANDRP).

87. Saunders, B., Sim, J., Kingstone, T., Baker, S., Waterfield, J., Bartlam, B., & Jinks, C., (2018). Saturation in qualitative research: Exploring its conceptualization and operationalization. *Quality & Quantity, 52*(4), 1893–1907.

88. Semenov, V. A., & Bengtsson, L., (2002). Secular trends in daily precipitation characteristics: Greenhouse gas simulation with a coupled AOGCM. *Climate Dyn., 19*, 123–140.

89. Shimizu, K., Short, D. A., & Kedem, B., (1993). Single and double-threshold methods for estimating the variance of area rain rate. *J. Meteor. Soc. Japan, 71*, 673–683.

90. Short, D. A., Shimizu, K., & Kedem, B., (1993a). Optimal thresholds for the estimation of area rain rate moments by the threshold method. *J. Appl. Meteor., 32*, 182–192.

91. Short, D., Wolff, B., Rosenfeld, D., & Atlas, D., (1993b). A study of the threshold method utilizing rain gauge data. *J. Appl. Meteor., 32*, 1379–1387.

92. Shry, C., & Reiley, E., (2010). *Introductory Horticulture: Cengage Learning* (Vol. 1, pp. 1–24). New York.

93. Soden, B. J., Wetherald, R. T., Stenchikov, G. L., & Robock, A., (2002). Global cooling after the eruption of Mount Pinatubo: A test of climate feedback by water vapor. *Science, 296*, 727–730.

94. Sorooshian, S., Gao, X., Hsu, K., Maddox, R. A., Hong, Y., Gupta, H. V., & Imam, B., (2002). Diurnal variability of tropical rainfall retrieved from combined GOES and TRMM satellite information. *J. Climate, 15*, 983–1001.

95. Sukovich, E. M., Kingsmill, D. E., & Yuter, S. E., (2009). Variability of graupel and snow observed in tropical oceanic convection by aircraft during TRMM KWAJEX. *Journal of Applied Meteorology and Climatology*, *48*(2), 185–198.

96. Suppiah, R., & Hennessy, K. J., (1996). Trends in the intensity and frequency of heavy rainfall in tropical Australia and links with the Southern oscillation. *Aust. Meteor. Mag., 45*, 1–17.

97. Tözeren, A., (2000). Laws of motion: Snowflakes, airborne balls, pendulums. *Human Body Dynamics: Classical Mechanics and Human Movement*, *1*, 30–55.

98. Trenberth, K. E., & Caron, J. M., (2000). The Southern oscillation revisited: Sea level pressures, surface temperatures and precipitation. *J. Climate, 13*, 4358–4365.

99. Trenberth, K. E., & Guillemot, C. J., (1994). The total mass of the atmosphere. *J. Geophys. Res., 99*, 23 079–23 088.

100. Trenberth, K. E., & Guillemot, C. J., (1996). Physical processes involved in the 1988 drought and 1993 floods in North America. *J. Climate, 9*, 1288–1298.

101. Trenberth, K. E., & Guillemot, C. J., (1998). Evaluation of the atmospheric moisture and hydrological cycle in the NCEP/NCAR reanalyses. *Climate Dyn., 14*, 213–231.

102. Trenberth, K. E., & Hoar, T. J., (1996). The 1990–1995 El Niño-Southern Oscillation event: Longest on record. *Geophys. Res. Lett., 23*, 57–60.

103. Trenberth, K. E., & Hurrell, J. W., (1994). Decadal atmosphere-ocean variations in the Pacific. *Climate Dyn., 9*, 303–319.

104. Trenberth, K. E., (1998). Atmospheric moisture residence times and cycling: Implications for rainfall rates with climate change. *Climatic Change, 39*, 667–694.

105. Trenberth, K. E., (1999a). Conceptual framework for changes of extremes of the hydrological cycle with climate change. *Climatic Change, 42*, 327–339.

106. Trenberth, K. E., (1999b). Atmospheric moisture recycling: Role of advection and local evaporation. *J. Climate, 12*, 1368– 1381.

107. Trier, S. B., & Parsons, D. B., (1993). Evolution of environmental conditions preceding the development of a nocturnal mesoscale convective complex. *Mon. Wea. Rev., 121*, 1078–1098.

108. Turtle, E. P., Del Genio, A. D., Barbara, J. M., Perry, J. E., Schaller, E. L., McEwen, A. S., & Ray, T. L., (2011). Seasonal changes in Titan's meteorology. *Geophysical Research Letters, 38*(3), 1–23.

109. Unsworth, M. H., & Wilshaw, J. C., (1989). Wet, occult, and dry deposition of pollutants on forests. *Agricultural and Forest Meteorology, 47*(2–4), 221–238.

110. Vivoda, J. M., Harmon, A. C., Babulal, G. M., & Zikmund-Fisher, B. J., (2018). E-hail (rideshare) knowledge, use, reliance, and future expectations among older adults. *Transportation Research Part F: Traffic Psychology and Behavior, 55*, 426–434.

111. Vuille, M., Bradley, R. S., Werner, M., Healy, R., & Keimig, F., (2003). Modeling $\delta^{18}O$ in precipitation over the tropical Americas. Part I: Interannual variability and climatic controls. *J. Geophys. Res., 108*, 4174, doi: 10.1029/2001JD002038.

112. Wakimoto, R. M., (1985). Forecasting dry microburst activity over the high plains. *Monthly Weather Review, 113*(7), 1131–1143.

113. Wallace, J. M., Zhang, Y., & Bajuk, L., (1996). Interpretation of interdecadal trends in Northern Hemisphere surface air temperature. *J. Climate, 9*, 249– 259.

114. Wang, J., Cole, H. L., Carlson, D. J., Miller, E. R., Beierle, K., Paukkunen, A., & Lane, T. K., (2002). Corrections of humidity measurement errors from the Vaisala RS80 radiosonde—Application to TOGA COARE data. *J. Atmos. Oceanic Technol., 19*, 981–1002.

115. Wilby, R. L., & Wigley, T. M. L., (2002). Future changes in the distribution of daily precipitation totals across North America. *Geophys. Res. Lett., 29*, 1135, doi: 10.1029/2001GL013048.

116. Wilson, J. W., & Schreiber, W. E., (1986). Initiation of convective storms at radar-observed boundary-layer convergence lines. *Mon. Wea. Rev., 114*, 2516–2536.

117. Wright, W. E., Long, A., Comrie, A. C., Leavitt, S. W., Cavazos, T., & Eastoe, C., (2001). Monsoonal moisture sources revealed using temperature, precipitation, and precipitation stable isotope timeseries. *Geophys. Res. Lett., 28*, 787–790.

118. Xie, P., & Arkin, P. A., (1997). Global precipitation: A 17-year monthly analysis based on gauge observations, satellite estimates, and numerical model outputs. *Bull. Amer. Meteor. Soc., 78*, 2539–2558.

119. Zhai, P., & Eskridge, R. E., (1997). Atmospheric water vapor over China. *J. Climate, 10*, 2643–2652.

Chapter 4

Role of Evaporation in Hydrometeorology

CONTENTS

4.1. INTRODUCTION

As previously stated, precipitation in the state of snow and rain is the main cause of water on the earth's crust. Evaporation and transpiration lose a significant amount of this water as vapor to the atmosphere. For instance, of the 1170 millimeters of water that India receives as precipitation each year, 597 millimeters (51%) is transferred to the sky via evaporation and transpiration, leaving only the remaining for subsurface formations, soils, and streams. In the United States, around 70% of the 750.0 mm of average annual precipitation is evaporated, lowering the total water gain by roughly three-quarters. These figures demonstrate the significance of evaporation in terms of water resource conservation and development. Evaporation is even more crucial in nations wherever surface water conservation is vital across a huge area (Meyer, 1915; Bowen, 1926).

Evaporation happens when water evaporates from the planet's land and water bodies. While transpiration happens when water evaporates from leaves via their stomata of plants. Somehow these two methods are distinct, they can be treated together in hydrometeorology because the mechanisms of force that produce transpiration and evaporation are regulated by analogous meteorological parameters like soil moisture, temperature, humidity, radiation, wind, sunshine, air pressure, and so forth. Evapotranspiration (ET) is the name given to this coupled process. Evaporation is important for estimating losses of water from lakes and reservoirs. It is important for crop growth modeling in agriculture (Rohwer, 1931; Penman et al., 1956).

4.2. PHYSICS OF EVAPORATION

Whenever there is an unrestricted water surface, such as in a pond or on bare land, evaporation takes place. The transfer of water from a liquid to a vapor or gaseous phase is known as evaporation. Evaporation can only happen if there is water available. Evaporation necessitates a significant amount of heat energy. One gram of water, for instance, requires around 580 calories of heat energy to evaporate. Heat sources include solar energy, heat stored from water, heat from earth ground, and sensible heat from the air itself (Zhang et al., 2012; Spence et al., 2013; Ramamurthy and Bou-Zeid, 2014).

Consider a pond full of water to best explain evaporation physics. Water has a lot of water molecules in it; hence it has a lot of mass. The differing velocities of these molecules are mostly determined by temperature. If some atoms in a body of water obtain sufficient kinetic energy, it moves away from the water's surface into the overspreading air and remain as vapor in the air as a result of a modest temperature rise. Water droplets in the overlying air are similarly in motion, but a few of them pierce the water's surface in the same way. Between the evaporating water surface and the surrounding above it, there is a constant exchange of water molecules occur. Evaporation occurs when the fleeing molecules outnumber the molecules returning. When the number of returning molecules surpasses the number of outgoing molecules, water vapor condenses (Saito et al., 2006; Pearlmutter et al., 2009; Zhang et al., 2010).

4.3. EVAPORATION AS A VAPOR FLOW NET RATE

Evaporation was originally recognized as a net rate of vapor transfer by John Dalton in 1802. A thin layer of air with the same temperature as the water is close to the water's surface. In a thin layer, the velocity of the departing molecules creates pressure. The vapor pressure refers to the pressure of hydrated vapor. The shallow layer may rapidly shift inundated with vapor at the equivalent temperature as the ocean. The air temperature and humidity just over the water surface or shallow layer determine its vapor pressure. Evaporation will happen if the real vapor pressure of the overspreading air is lower than that of the water's surface or shallow layer. The disparity among the vapor pressure (ea) in the overlying air and the vapor pressure represented by 'ew' at the surface, as Dalton's law, determines the flow of evaporation from the water surface (Leary and Houze, 1979; Lee et al., 2012). Mathematically,

$$E = C\,(e_w - e_a) \qquad (1)$$

where; C represents a coefficient depending on the velocity of the wind and many factors linked to evaporation; e_a represent the exact vapor pressure within air overlying (millimeters of mercury); e_w represent the saturated vapor pressure related to the water temperature or of the shallow layer (mm of mercury), and E represents the evaporation rate (mm per day).

4.4. FACTORS AFFECTING EVAPORATION

Both non-meteorological and meteorological factors affect evaporation from the surface of the water. Atmospheric pressure, humidity, solar radiation, wind, and water and air temperature are all meteorological factors. The depth of the water, the quality of the evaporating surface, and the area of the water surface are all non-meteorological factors. The following sections provide a brief overview of these factors (Monteith, 1981; Plawsky et al., 2008). The amount of heat energy accessible determines how quickly water evaporates from the substrate. Radiation from the sun is the most significant energy resource and establishes wide temperature boundaries. In the tropics, solar radiation is abundant, resulting in high evaporation (Singh and Narayan, 1990; Zang et al., 2019).

Evaporation tends to increase when the water temperature and air surface is high because air and water temperature are associated with solar radiation. The gradient of vapor pressure is controlled by the temperature of water's surface. Because the air's ability to withstand water vapor rises as the overlying air temperature rises, evaporation will occur quickly. As a result, evaporation is greater in the summer or in hotter countries than in the winter or in colder countries (Eames et al., 1997; Buckley et al., 2017).

Evaporation is influenced by humidity. It is a measurement of how much water vapor is already in the air, and thus identification of its ability to incorporate vapor due to the evaporation from the water surface. Evaporation can be reduced if the air humidity is higher. Because water vapor travels from a higher to a lower vapor content throughout evaporation. As the movement rate is determined by the disparity in vapor amount or the air humidity gradient. As a result, a sharp humidity gradient apart from the water surface is critical for rapid evaporation (Barrett and Clement, 1992; Asteman et al., 2000).

Evaporation is influenced by wind speed as well. Evaporation is the process of removing vapor from a water surface and transporting it to a shallow layer of air closer to the surface. Evaporation comes to a halt when this shallow layer is saturated. If there is a higher disturbance or if the speed of wind directly over the water surface is high enough, the saturated air layer is simply replaced by the wind, allowing for high evaporation. Modest wind speeds incline to keep the gradient of humidity high and the rates of evaporation large (Milly, 1984).

4.5. EVAPORATION MEASUREMENT

Uncovering a container or pan of water in the air can be used to calculate the amount of evaporation, which is typically described in millimeters. The pan is loaded up to some level with the water, and the decline in water level is calculated over a set period (typically a 24-hour). According to research, evaporation from a large water body, like a reservoir or a lake, differs from evaporation from a little pan. A proper pan coefficient is used to reconfigure the evaporation through a small pan to achieve the evaporation from a water body. Different designers created different pan structures, and different pan coefficient values were assigned. Below is a list of some of the most essential pans (Ritchie and Johson, 1990; Borchman et al., 2009).

4.5.1. Class A Pan-Evaporimeter of U.S. Weather Bureau

This pan has a diameter of 1210 millimeters (4 feet), is 255 millimeters (10-inch) deep, and is positioned on a wooden board 150 millimeters (6-inch) well above ground. Water is poured into the pan to a depth of 180 to 200 millimeters. With a hook scale in a stilling well, they fall in water level over individual day is monitored. Usually, the pan is formed of a galvanized sheet of iron that has not been painted. Where other meteorological tools are maintained, the evaporimeter is assembled. With an overall average of 0.7, the pan coefficient ranges between 0.6 and 0.8. This evaporimeter is used to measure evaporation by weather services all over the world. Every day, the weather services report on evaporation at various locations. The class A pan evaporimeter from the United States Weather Bureau is illustrated in Figure 4.1 (Penman and Keen, 1940; Penman et al., 1967).

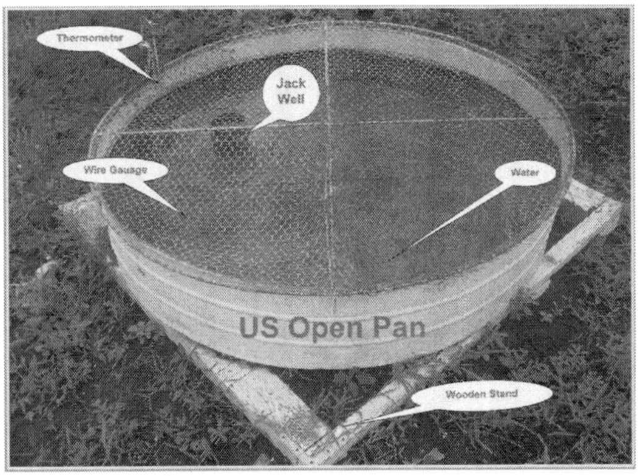

Figure 4.1. Class A open pan evaporimeter of U.S. Weather Bureau.

Source: *Jaybhaye et al. (2013).*

4.5.2. Colorado Sunken Pan

This pan is composed of an unpainted galvanized iron sheet and its geometry is 460 mm deep by 920 mm square. The pan is sunk so that barely 100–150 mm of it protrudes just above the ground surface. The level of water in the pan is kept about level with the ground. A point gauge is used to quantify evaporation. The submerged pan's key advantage is that it has similar wind and radiation properties of a lake. The pan coefficient ranges from 0.75 to 0.86, with a mean of 0.78.

4.5.3. Floating Pan of U.S. Geological Survey

It is a 450 mm deep, 900 mm square pan set floating across a lake and supported with drum floats in a raft. With the walls of the pan extending 75 mm over the lake, the level of water in the pan is maintained at a similar level as that of the lake. Coefficient of the pan is commonly assumed to be 0.80. The pan resembles the qualities of a lake in appearance (Harris et al., 1916; Yan et al., 2019).

Table 4.1 shows the recommendations of the World Meteorological Organization (WMO) for the limited evaporation stations number required for calculation of evaporation in various geographical locations.

Table 4.1. Criteria for Evaporation Stations Set by the World Meteorological Organization

Area	Lower Density km²/Station
1. Arid area	30,000.0
2. Humid temperate climate	50,000.0
3. Cold area	100,000.0

4.6. INDIA MEASUREMENT OF EVAPORATION

Evaporation is calculated by utilizing class A pan evaporimeters from the US Weather Bureau in India. The India Meteorological Department monitors evaporation at over 200 stations around the country regularly. The Meteorological Department of India (IMD) reported yearly and monthly and mean evaporation figures for 30 stations within India in 1980 (Jaswal et al., 2014). Table 4.2 shows the seasonally daily mean evaporation for several stations, and Figure 4.2 shows the spatial change of average annual evaporation across the Indian territory. There are some data published that are preserved by state government agricultural and irrigation departments. Evaporation varies in between 500 and 1500 mm from the northeast to the dry parts of western Gujarat and Rajasthan, as seen in Figure 4.2. In semi-arid and arid regions, the average evaporation rate usually exceeds the region's local rainfall. Consequently, a substantial amount of water is lost (Shuttleworth et al., 1988; Schindler et al., 2010).

It is stated that obtaining evaporation estimates from big water bodies such as reservoirs and lakes, is critical for expanding water storage and assuring effective water resources management. Evaporation from huge water bodies, such as reservoirs or lakes, is not reflected by pan measurements. To calculate evaporation from reservoirs and lakes, a variety of approaches have been devised. The following section will go over these techniques (Yanful et al., 2003).

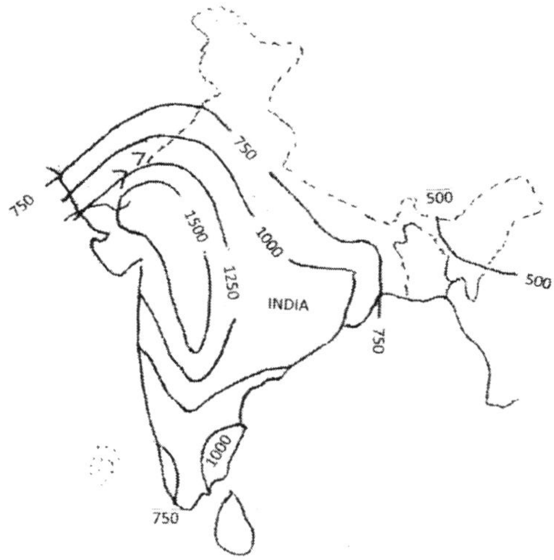

Figure 4.2. In India, Spatial variation of mean evaporation (mm).

Source: *https://link.springer.com/book/10.1007%2F978-1-4020-9844-4.*

4.7. EVAPORATION ESTIMATION USING INDIRECT METHODS

When direct measurements for reservoirs and lakes and are unavailable, the following approaches are used to calculate the quantity of evaporation (Ali et al., 2008; Patel et al., 2016):

- Water budget techniques;
- Pan coefficients usage;
- Technique of vapor flow;
- Technique of energy balance;
- Union of vapor flow and energy techniques;
- Empirical formulae.

Penman (1963), Wieisner (1970), Oliver (1961), and Chang (1968) are only a few of the references that show the detailed procedures. These techniques are briefly covered below.

Table 4.2. Seasonally Averaged Daily Evaporation in Millimeters, as observed by a Class A Pan Evaporimeter of the US Weather Bureau

Station	Winter	Hot Weather	SW-Monsoon	Post Monsoon	Annual
	December-February	**March-May**	**June-September**	**October-November**	
Agartala station	3.20	6.50	4.50	3.20	4.40
Ahmedabad station	5.80	11.20	6.70	6.30	7.50
Allahabad station	2.60	7.20	5.50	3.50	4.70
Agra station	3.50	10.50	8.00	4.30	6.60
Bangalore station	4.80	7.20	4.10	3.70	4.90
Bikaner station	3.20	9.50	9.50	5.40	6.90
Mumbai station	3.70	5.60	3.50	3.90	4.20
Kolkata station	3.00	5.70	4.10	3.60	4.10
Hyderabad station	6.20	10.00	6.70	5.30	7.00
Jodhpur station	2.10	5.60	4.40	2.80	3.70
Kota station	4.60	12.40	8.80	5.60	7.80
Chennai station	4.20	6.30	6.80	3.70	5.30
New Delhi station	3.70	8.70	7.10	5.40	6.20
Patna station	3.00	9.10	5.70	3.30	5.30
Roorkee station	2.50	8.20	6.10	3.40	5.10

Trivandrum station	4.10	4.40	3.80	3.70	4.00
Pune station	5.20	10.30	4.70	4.90	6.30

4.7.1. Technique of Water Budget

Water budget technique for calculating evaporation from the reservoir or a lake is dependent on a balance of storage changes, outflows, and inflows (Chow, 1964; Linsley et al., 1975). This entails formulating the equation of the lake's hydrologic continuity and calculating evaporation based on certain parameters. The t equation of water budge is expressed as follows when using lake daily average values:

$$P + R_1 + R_g = R_0 + R_s + E + T \pm S \quad (2)$$

or

$$E = P + R_1 + R_g - R_0 - R_s - T \pm S \quad (3)$$

where; R_1 is the lake surface inflow on daily basis; Rg is the groundwater inflow on daily basis; R_0 is the lake surface outflow on daily basis; Rs is the seepage outflow on daily basis; E is the lake evaporation on daily basis, daily transpiration; and S is the per day change in lake storage; and P is the daily precipitation.

The net seepage outflow via the surface of the ground is important to deal with, therefore Rg – Rs = 0. T stands for negligible transpiration losses. Eqn. (3) is transformed into

$$E = P + R_1 - R_0 \pm S \quad (4)$$

(+ value for increased storage, – value for decreased storage). In Eqn. (4) all the terms are in-depth (mm) or volume (m^3) units for a specific period. *E* can be determined if other parameters are known. If the time unit is maintained large, such as a week or longer, the values of E can be more accurate. Although the water budget approach has the clear advantage of having easy in theory, it cannot be anticipated to yield fairly accurate findings due to different uncertainties and the possibility of inaccuracies in measured parameters. Controlled experiments for Lake Hefner in the United States, on the other hand, have yielded fairly correct results utilizing this method (Eberbach et al., 2011).

4.7.2. Energy Balance Techniques

This technique work on the principles of heat energy conservation and evaporative cooling. In the evaporation process, heat is wasted and the loss of heat must match the gain of heat. The quantity of evaporation may be calculated if overall heat energy approaching the water surface and energy leaving the water surface and the stored energy in the water are known throughout a specific period (Moghaddamnia et al., 2009; Shirsath and Singh, 2010).

Consider Figure 4.3, which depicts a body of water. It is possible to account for the heat sources that contribute to the thermal balance at the surface of the water. As a result, for a given amount of time, the equation of energy budget for the surface evaporating is expressed as:

$$(1 - r)\, H_c = H_b + LE + H + H_a + G + H_s \qquad (5)$$

$$(1 - r)\, H_c - H_b = LE + H + H_a + G + H_s \qquad (6)$$

$$Q_n = LE + H + H_a + G + H_s \qquad (7)$$

where; Ha is the heat flow into or out from the water by water inflow or outflow or; Hc represent incident of solar radiation to the water surface; rHc is the reflected solar radiation from the albedo(r) water surface; Hc (1 – r) is the solar radiation into the water body; r is albedo; H is the air to water transfer of sensible heat or in the opposite direction; Hb is the water body outgoing radiation; LE is the heat energy utilize in evaporation; and L is the evaporation latent heat; Hs is the stored heat in the body of water; G is the heat flux inflow to the ground; and Qn is the received net heat energy by the water surface.

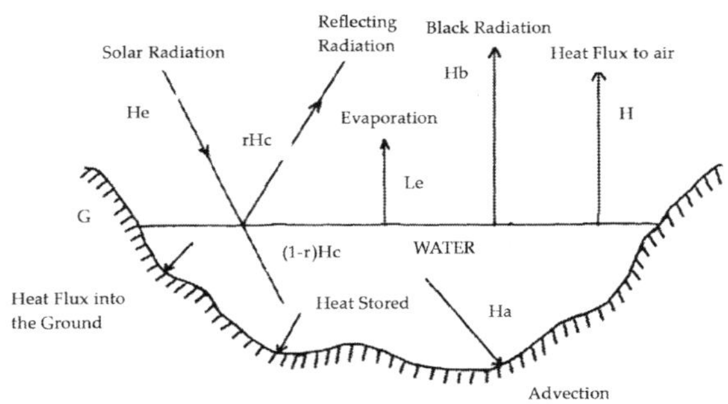

Figure 4.3. Diagram of the energy balance.

Source: *https://link.springer.com/book/10.1007%2F978-1-4020-9844-4.*

The heat flux of ground(G) is minimal for several land surfaces and it can be ignored (Table 4.3). In addition, if the period is short, the terms Ha and Hs are insignificant. As a result, Eqn. (7) takes on a new meaning as:

$$Q_n = LE + H \qquad (8)$$

Table 4.3. Albedo Values for Some Surfaces

Surface	**Albedo**
Water at 0 to 60° latitude (N)	<0.080
Water at 60 to 90° latitude (N)	<0.100
Snow	0.80
Surface of Earth's	0.150
Land of wood	0.140
Tropical forest	0.130
Natural grassland	0.20

4.7.3. Method of Vapor Flow

The quantity of water vapor pass from the surface of the water to the atmosphere is the basis for the vapor flow technique of evaporation. As previously mentioned, Dalton (1802) demonstrated that the evaporation rate is determined by the gradient of vapor pressure among the air above and evaporating surface, which is defined as the distinction among the saturated vapor pressure (e_w) at the surface temperature of the water and the overlying air vapor pressure (e_a), as well as a vapor removal measure which should be a wind speed function (Chang et al., 2010):

$$E = c\,f(u)\,(e_w - e_a) \quad (9)$$

where; E represents the evaporation rate (mm per day); and f (u) is a wind function. Eqn. (10) is expressed in a more generic form as:

$$E = c(e_w - e_a)\,(a + bu) \qquad (10)$$

where; u signifies an average speed of wind at a reasonable height; and a, b, and c are constant. The terms e_a and e_w have their typical meanings, which have already been discussed. If the wind velocity and distribution of vapor are known, the vapor flow method could be employed for the calculation of evaporation.

4.7.4. Pan Coefficients Usage

Pan evaporimeter measurements of evaporation may not always reflect the evaporation rate from a reservoir. As a result, these data cannot be used to determine the depth of a reservoir or lake. This is because a lake is larger in deepness and area or may be subjected to various wind rates than a pan. In contrast to the reservoir or lake, the little water volume within the metallic pan is severely impacted by changes in temperature caused by solar radiation. As a result, lake evaporation is far lower than evaporation from a measuring pan. As a result, data of pan evaporation must be modified by using a pan coefficient to obtain true evaporation from reservoirs and lakes. A coefficient is defined as the ratio between the evaporation of the lake and pan.

$$\text{Pan coefficient} = \frac{\textit{Evaporation from the lake}}{\textit{Evaporation from pan evaporimeter}}$$

Pan coefficient is measured by determining the theoretical evaporation of the lake and correlating it to the pan. Yearly lake evaporation to pan evaporation ratio is stable from region to region and year to year, although monthly ratios vary a lot. Pan coefficients vary between 0.67 and 0.82 yearly, with just an average of 0.7. As a result, yearly, the evaporation of the lake is roughly 0.7 times the evaporation of the pan. The evaporation rate from exposed wet soil is around 0.60 times that of the pan. Khan and Bohra in 1990 found an annual pan coefficient of 0.67 for the western Rajasthan region of India. Table 4.4 shows the figure of pan coefficients for various pans.

Table 4.4. Pan Coefficients' Values

Pan Types	**Mean Value**	**Range**
1. U.S. Weather Bureau Class A	0.70	0.600–0.800
2. Colorado Sunken	0.78	0.750–0.860
3. USGS Floating Pan	0.80	0.700–0.820

REFERENCES

1. Ali, S., Ghosh, N. C., & Singh, R., (2008). Evaluating best evaporation estimate model for water surface evaporation in semi-arid region, India. Hydrological processes: *An International Journal, 22*(8), 1093–1106.
2. Asteman, H., Svensson, J. E., Norell, M., & Johansson, L. G., (2000). Influence of water vapor and flow rate on the high-temperature oxidation of 304L; Effect of chromium oxide hydroxide evaporation. *Oxidation of Metals, 54*(1), 11–26.
3. Barrett, J., & Clement, C., (1992). Kinetic evaporation and condensation rates and their coefficients. *Journal of Colloid and Interface Science, 150*(2), 352–364.
4. Borchman, D., Foulks, G. N., Yappert, M. C., Mathews, J., Leake, K., & Bell, J., (2009). Factors affecting evaporation rates of tear film components measured *in vitro*. *Eye & Contact Lens, 35*(1), 32–37.
5. Bowen, I. S., (1926). The ratio of heat losses by conduction and by evaporation from any water surface. *Phys. Rev.*, *27*, pp. 779–789.
6. Buckley, T. N., John, G. P., Scoffoni, C., & Sack, L., (2017). The sites of evaporation within leaves. *Plant Physiology, 173*(3), 1763–1782.
7. Chang, F. J., Chang, L. C., Kao, H. S., & Wu, G. R., (2010). Assessing the effort of meteorological variables for evaporation estimation by self-organizing map neural network. *Journal of Hydrology, 384*(1–2), 118–129.
8. Chang, J. H., (1968). *Climate and Agriculture* (Vol. 1, pp. 1–23). Aldine, Chicago.
9. Chow, V. T., (1964). *Handbook of Applied Hydrology* (Vol. 1, pp. 1–20). McGraw Hill, New York, NYS.
10. Dalton, J., (1802). Experimental essays on the constitution of mixed gases on the force of steam or vapor from water and other liquids in different temperatures, both in a *Torricelian vacuum* and in air on evaporation and on the expansion of gases by heat. *Manchester Lu. Phil. Soc. Mem. Proc., 5*, 536–602.
11. Eames, I. W., Marr, N. J., & Sabir, H., (1997). The evaporation coefficient of water: A review. *International Journal of Heat and Mass Transfer, 40*(12), 2963–2973.
12. Eberbach, P. L., Humphreys, E., & Kukal, S. S., (2011). The effect of rice straw mulch on evapotranspiration, transpiration, and soil evaporation of irrigated wheat in Punjab, India. *Agricultural Water*

Management, 98(12), 1847–1855.

13. Harris, F. S., Robinson, J. S., & Agronomy, F., (1916). Factors affecting the evaporation of moisture from the soil. *J. Agr. Res., 7*(10).
14. Jaswal, A. K., Narkhede, N. M., & Rachel, S., (2014). Atmospheric data collection, processing, and database management in India Meteorological Department. In: *Proc. Indian Natl. Sci. Acad.* (Vol. 80, pp. 697–704).
15. Jaybhaye, P., Raundal, P., & Andhale, P., (2013). *Handbook of Agrometeorology* (pp. 177–180). Contenental Prakashan, Vijaya Nagar, Pune – 411 030, Maharashtra, India.
16. Leary, C. A., & Houze Jr, R. A., (1979). Melting and evaporation of hydrometeors in precipitation from the anvil clouds of deep tropical convection. *Journal of Atmospheric Sciences, 36*(4), 669–679.
17. Lee, J. E., Risi, C., Fung, I., Worden, J., Scheepmaker, R. A., Lintner, B., & Frankenberg, C., (2012). Asian monsoon hydrometeorology from TES and SCIAMACHY water vapor isotope measurements and LMDZ simulations: Implications for speleothem climate record interpretation. *Journal of Geophysical Research: Atmospheres, 117*(D15), 1–25.
18. Linsley, R. K., Kohler, M. A., & Paulhus, J. L. H., (1975). *Hydrology for Engineers* (2nd edn., Vol. 1, pp. 340). New York, McGraw-Hill.
19. Meyer, A. F., (1915). Computing runoff from rainfall and other physical data. *Trans. ASCE, 79*, pp. 1056–1115.
20. Milly, P. C. D., (1984). A simulation analysis of thermal effects on evaporation from soil. *Water Resources Research, 20*(8), 1087–1098.
21. Moghaddamnia, A., Gousheh, M. G., Piri, J., Amin, S., & Han, D., (2009). Evaporation estimation using artificial neural networks and adaptive neuro-fuzzy inference system techniques. *Advances in Water Resources, 32*(1), 88–97.
22. Monteith, J. L., (1981). Evaporation and surface temperature. *Quarterly Journal of the Royal Meteorological Society, 107*(451), 1–27.
23. Oliver, H., (1961). *Irrigation and Climate* (Vol. 1, pp. 1–20). Edward Arnold, London.
24. Patel, J. N., & Majmundar, B. P., (2016). Development of evaporation estimation methods for a reservoir in Gujarat, India. *Journal-American Water Works Association, 108*(9), E489–E500.

25. Pearlmutter, D., Krüger, E. L., & Berliner, P., (2009). The role of evaporation in the energy balance of an open-air scaled urban surface. *International Journal of Climatology: A Journal of the Royal Meteorological Society, 29*(6), 911–920.

26. Penman, H. L. et al., (1956). Discussion of evaporation. *Netherlands J. of Agr. Sci., 4*, 87–97.

27. Penman, H. L., & Keen, B. A., (1940). Meteorological and soil factors affecting evaporation from fallow soil. *Quarterly Journal of the Royal Meteorological Society, 66*(287), 401–410.

28. Penman, H. L., (1948). Natural evaporation a review. *J. of Hydrol., 13*, 1–21.

29. Penman, H. L., (1963). *Vegetation and Hydrology* (Vol. 53, p. 124). Commonwealth Bureau of Soils Harpenden, Tech. communication.

30. Penman, H. L., Angus, D. E., & Van Bavel, C. H. M., (1967). Microclimatic factors affecting evaporation and transpiration. *Irrigation of Agricultural Lands, 11*, 481–505.

31. Plawsky, J. L., Ojha, M., Chatterjee, A., & Wayner, Jr. P. C., (2008). Review of the effects of surface topography, surface chemistry, and fluid physics on evaporation at the contact line. *Chemical Engineering Communications, 196*(5), 658–696.

32. Ramamurthy, P., & Bou-Zeid, E., (2014). Contribution of impervious surfaces to urban evaporation. *Water Resources Research, 50*(4), 2889–2902.

33. Ritchie, J. T., & Johson, B. S., (1990). Soil and plant factors affecting evaporation. *Agronomy, 30*, 363–390.

34. Rohwer, C., (1931). *Evaporation from Free Water Surface* (Vol. 1, pp. 260–271). US Dept. Agr: Tech. Bull.

35. Saito, K., Yasunari, T., & Takata, K., (2006). Relative roles of large-scale orography and land surface processes in the global hydroclimate. Part II: Impacts on hydroclimate over Eurasia. *Journal of Hydrometeorology, 7*(4), 642–659.

36. Schindler, U., Durner, W., von Unold, G., & Müller, L., (2010). Evaporation method for measuring unsaturated hydraulic properties of soils: Extending the measurement range. *Soil Science Society of America Journal, 74*(4), 1071–1083.

37. Shirsath, P. B., & Singh, A. K., (2010). A comparative study of daily pan evaporation estimation using ANN, regression, and climate-based models. *Water Resources Management, 24*(8), 1571–1581.

38. Shuttleworth, W. J., Gash, J. H. C., Lloyd, C. R., McNeill, D. D., Moore, C. J., & Wallace, J. S., (1988). An integrated micrometeorological system for evaporation measurement. *Agricultural and Forest Meteorology, 43*(3–4), 295–317.
39. Singh, R. K., & Narayan, J., (1990). Pulsed-laser evaporation technique for deposition of thin films: Physics and theoretical model. *Physical Review B, 41*(13), 8843.
40. Spence, C., Blanken, P. D., Lenters, J. D., & Hedstrom, N., (2013). The importance of spring and autumn atmospheric conditions for the evaporation regime of Lake Superior. *Journal of Hydrometeorology, 14*(5), 1647–1658.
41. Yan, Z., Wang, S., Ma, D., Liu, B., Lin, H., & Li, S., (2019). Meteorological factors affecting pan evaporation in the Haihe River Basin, China. *Water, 11*(2), 317.
42. Yanful, E. K., Mousavi, S. M., & Yang, M., (2003). Modeling and measurement of evaporation in moisture-retaining soil covers. *Advances in Environmental Research, 7*(4), 783–801.
43. Zang, D., Tarafdar, S., Tarasevich, Y. Y., Choudhury, M. D., & Dutta, T., (2019). Evaporation of a droplet: From physics to applications. *Physics Reports, 804*, 1–56.
44. Zhang, Y., Leuning, R., Chiew, F. H., Wang, E., Zhang, L., Liu, C., & Jung, M., (2012). Decadal trends in evaporation from global energy and water balances. *Journal of Hydrometeorology, 13*(1), 379–391.
45. Zhang, Y., Leuning, R., Hutley, L. B., Beringer, J., McHugh, I., & Walker, J. P., (2010). Using long-term water balances to parameterize surface conductance and calculate evaporation at 0.05 spatial resolution. *Water Resources Research, 46*(5).

Chapter 5

Hurricanes and Tropical Storms

CONTENTS

5.1. INTRODUCTION

Severe cyclonic storms and depressions are all tropical disturbances with less than 63 knots of wind speeds. Between latitudes of 7° and 15°, stronger cyclonic storms above 63 knots wind speed spiral around tropical centers with a low-pressure center. Cyclones in India, and hurricanes in the Caribbean and North America, as well as typhoons in Japan, are examples of such systems. Following their formation, they begin traveling westward at a speed of roughly 15 kilometers per hour across open ocean waters. Tropical cyclones bring thick clouds, turbulent waves, and a lot of rain. A well-developed storm poses a threat to nearby ships as well as the coastal area where it makes landfall. In its forward sector, the storm builds up a massive seawater sheet that inundates coastal regions which are low-lying, creating widespread destruction and death. Storm surge can reach 80 kilometers wide, also four meters deep, and acts as the most dangerous component of a hurricane. Storm surges are responsible for nine out of 10 hurricane deaths. Consider the historical tropical cyclone of November 1970, which hit Bangladesh due to storm surge, caused the death of approximately 300,000, whereas hurricane Mitch of October 1998, which hit Central America, killed over 11,000 people also cost millions of dollars in damage. Tropical cyclones bring severe and widespread rainfall for multiple days as they pass over land. In hydrology, the significance of tropical cyclones is based on the fact that they incase water quantities on land, that, for life, is necessary, as well as the destruction they cause to coastal areas. The Central United States receives a majority of its rain through hurricanes (Figure 5.1).

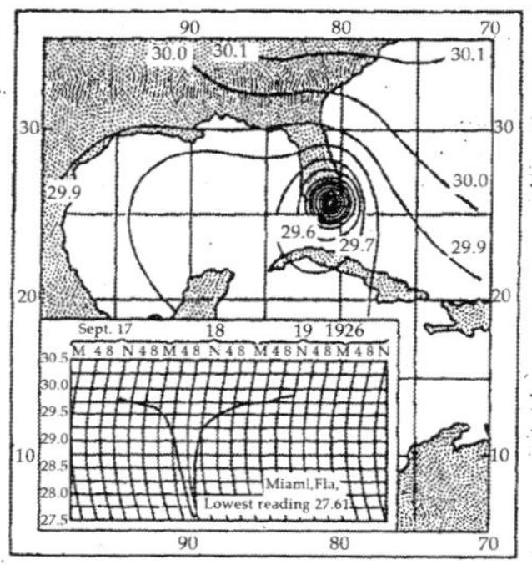

Figure 5.1. Hurricane, in Florida's southern tip, of September, 1926 (Petterssen, 1969).

Source: *https://www.amazon.com/Introduction-Meteorology-S-Petterssen/dp/0070497206.*

5.2. SAFFIR-SIMPSON HURRICANE SCALE

High winds, Storm surge, territorial rainfall, floods are some sources of damage caused by hurricane. As a result, an organization system which rates tropical hurricanes and storms, solely, on wind speed basis is insufficient in adequate representation of hurricanes, especially when talking of the repercussions.

Robert Simpson, Director of the National Hurricane Center (NHC) and a meteorologist, with Herbert Saffir, a consulting engineer in Florida, devised a ranking scale in 1970 to classify and assess hurricane damage potential. The Saffir Simpson classification combines coastal destruction potential and storm surge height in addition to central pressure and wind speed. This Scale ranks the hurricane intensity on a 1–5 scale based on storm surge height, air pressure, and wind speed. The scale and the damage grading criteria are shown in Table 5.1.

The Saffir-Simpson hurricane intensity scale is used in the United States to predict the possibility for damage to property as well as flooding along the coast as a result of a storm's landfall.

From June through November, on average, 8–9 Atlantic tropical storms form, counting the Caribbean Sea and the Gulf of Mexico. Only 2 to 3 of these are likely to become hurricanes, and only 4–5 of these are anticipated to reach category 3 Saffir-Simpson or higher intensity. Hurricanes of category 5 are extremely unusual, happening approximately every 100 years or so.

Table 5.1. Hurricane Scale (Saffir-Simpson)

Saffir-Simpson	**Wind**			**Central Pressure**		**Storm Surge Height**		**Damage**	**North American**
Category	**mph**	**m/s**	**Knots**	**Inches**	**mb**	**ft**	**m**		**Examples**
1	74–95	33–42	64–82	>28.94	>980	3–5	1.0–1.7	Minimal	Agnes, 1972
2	96–110	43–49	83–95	28.91–28.50	979–965	6–8	1.8–2.6	Moderate	Cleo, 1964
3	111–130	50–58	96–113	28.47–27.19	964–945	9–12	2.7–3.8	Extensive	Fran, 1996
4	131–155	59–69	114–135	27.88–27.17	944–920	13–18	3.9–5.6	Extreme	Andrew, 1992
5	>156	>70	>136	<27.12	<920	>18	>5.7	Catastrophic	Camille, 1969

5.3. SUPER TYPHOONS AND MAJOR HURRICANES

Wind speeds can sometimes be significantly higher than those seen within a hurricane. In addition, the destruction observed from hurricanes is not proportional to speed of wind. Instead, when the wind speed increases, the amount of damage done increases exponentially. The maximum sustained winds associated with hurricanes determine their categorization as major hurricanes. As a result, the NHC uses the term "major hurricane" to describe hurricanes with highest maintained surface winds (at least 96 knots) of 1 minute. On the Saffir-Simpson Scale, this corresponds to Category 3, 4, or 5. To get a consistent measurement, the World Meteorological Organization (WMO) recommends using a 10-minute average. The Joint Typhoon Warning Center (JTWC) and NHC of the United States of America, on the other hand, utilize an average time of 1-minute to calculate maintained winds. The United States' JTWC uses the term "super typhoon" to describe typhoons with highest maintained 1-minute surface winds with 130 knots at least. In the Atlantic Ocean, this is the comparable to a strong Saffir-Simpson 4 or 5 Category hurricane.

5.4. WIND ROTATION

Due to the rotating movement, from west to east, of the globe about its axis, low-pressure system containing winds, in northern hemisphere flow counter-clockwise and vice versa. The Coriolis force, apparent result of earth's rotation, is also sometimes known as the deviating force. This force pushes an object westward as it approaches the equator, straight-line path gets converted to curved path due to the rotation of the planet. Moving things in the northern hemisphere tend to deflect to the right direction to their line of motion, and vice versa. As a result, the Coriolis force acts to pull the winds rightward, in the Northern hemisphere and vice versa. In north of the equator, during formation of a low-pressure system, the air moves to low-from a high-pressure zone to restore equilibrium. A straight path of wind from high to low is not observed because of the earth's rotation. In the northern hemisphere, winds aimed at the core of an area of low pressure deflect to the right, resulting in circulation around the low-pressure area to be anticlockwise rather direct impingement from every direction.

A revolving cup anemometer is usually used to measure wind speed and is expressed in terms of meters per second (m/s), miles per hour (mph), kilometers per hour (km/hr), and knots (nautical miles per hour), the units are compared as follows:

1.1574 mph = 1 knot = 1.853 km/hr = 0.5148 m/s

1 mph = 0.864 knot = 1.609 km/hr = 0.447 m/s 1 mph = 0.864 knot = 1.609 km/hr = 0.447 m/s

1 km/hr = 0.538 knot = 0.621 mph = 0.278 m/s 1 km/hr = 0.538 knot = 0.621 mph = 0.278 m/s

5.5. FORMATION OF TROPICAL STORMS

On both sides of the equator, the semitropical high-pressure belts are around 30° apart. The tropical zone is the area of the world that lies between these two high-pressure belts. The southeast (SE) and northeast (NE) trade winds blow from these high-pressure areas towards the equator, where they merge at the ITCZ, a low-pressure zone. The ITCZ is an area of ascending motion and maximum cloudiness due to the convergence of trade winds. Because of the deceptive north-south passage of the sun among the tropics of Capricorn and tropics of Cancer the over the sequence of a year, this ITCZ has a considerable seasonal shift from summer to winter. As a result, during the northern summer, equator is to the north of the center of the meteorological systems of both hemispheres., while during the southern summer, it is to the south of the equator. Furthermore, the sun is directly above twice a year in tropical locations, and its beams are seldom extremely oblique. As a result, the air over hot oceans are often warm and moist. Warm air over tropical oceans with a great moisture content are substantially lighter than warm air with a low moisture content elsewhere. Low-pressure zones with high convective activity result from warm wet air. When the atmospheric pressure in an area of the ocean drops slightly, the surrounding air begins to move in a round or almost circular pattern around the center of low pressure, eventually spinning. Due to the rotating motion of the planet around its own axis from west to east, the wind surrounding like a low-pressure core cycle anticlockwise in the northern Hemisphere and circular in the southern Hemisphere. A tropical storm is the pressure and wind system that surrounds a low-pressure center. Tropical storms (Figure 5.2) have practically round isobars around the center of low pressure on the surface weather map.

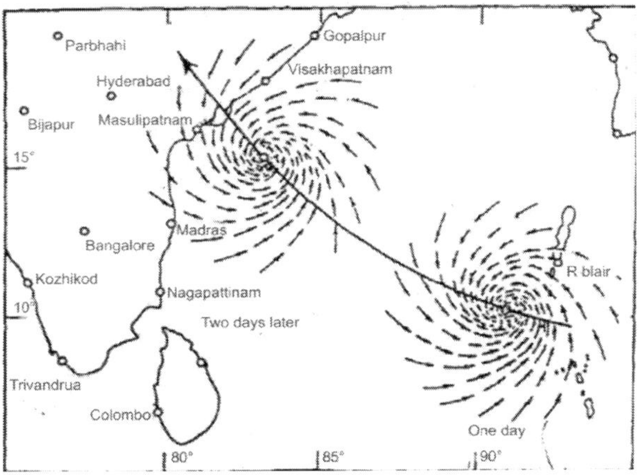

Figure 5.2. Tropical cyclone in the Bay of Bengal.

Source: *https://link.springer.com/chapter/10.1007/978-1-4020-9844-4_7.*

5.6. ORIGIN OF HURRICANES

During the northern summer, hurricanes in the Atlantic Ocean frequently begin near Africa. They are pushed westward through easterly trade winds, and once they meet the predominant easterly winds throughout North America, they shift northward. Lower tropospheric westward moving disturbances have been recognized as the seeding circulations for a major fraction of hurricanes across the North Atlantic Ocean since the 1930s. The disturbances that flow westward are known as easterly waves, and they originate in North Africa. The unpredictability of the African jet causes these easterly waves. These easterly waves flow westward over the Atlantic Ocean in the lower troposphere trade wind belt, causing cloudy weather in the places they affect. Through the trade wind belt, these waves normally start in April or May and last until October or November.

Waves have a wavelength of 2,000 to 2,500 km and a duration of 3 to 4 days (1,200 to 1,500 miles). The presence of a modest low-pressure core at the equatorial end of an easterly wave's trough is a distinguishing feature. Such an easterly wave-related center can sometimes evolve into a tropical depression, then a storm, and finally a hurricane. Each year, roughly 60 waves are produced over North Africa, although it seems that the number of waves made has no relation to the amount of hurricane activity in the Atlantic.

5.7. STRUCTURE OF HURRICANES

A hot cyclone is a massive rotating column of moist air with a diameter of 500–1000 km and a height of 10–15 km. The most exciting characteristic of this massive revolving mass of wet air is that it has a funnel-shaped funnel in the center, known as the cyclone's 'eye.' The eye has a diameter of around 20–30 kilometers. The air pressure is lowest here. The lowest pressure on record for hurricanes in the United States is 892 mb measured at Florida, on September 2, 1935, whereas the lowest pressure at the center of Indian cyclones is around 919 mb, reported at False Point on September 22, 1885. A satellite cloud image of the eye is the best way to see it. In the center of a vast area of bright clouds, the eye appears as a dark point. The eye is almost cloudless at the top, and the wind is either absent or extremely feeble. The walls of the funnel, on the other hand, represent a zone of extremely intense wind rotation (Tannehill, 1956). The top sketch of Figure 5.3 depicts the spinning air's direction; the middle sketch of Figure 5.3 depicts the hurricane's well-shaped essential eye, in which air movement is directed descending while air movement is directed upward on the sides. The bottom graphic depicts a part of the hurricane's funnel as it passed over Manila, Philippines, in 1882. The edge of the funnel is quite steep up to a height of 8 kilometers, then becomes gentler farther up.

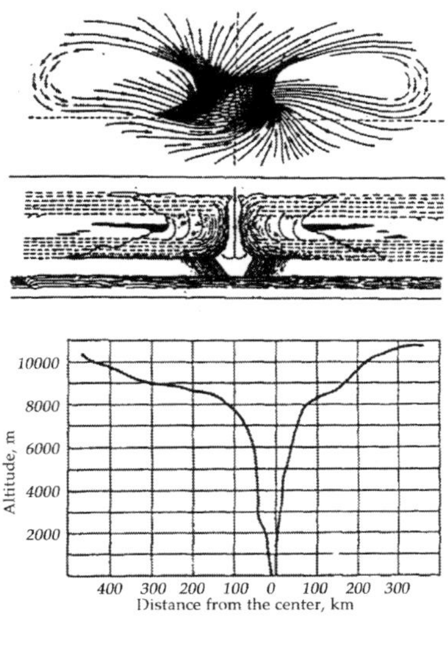

Figure 5.3. Structure of a hurricane (Tannerhill, 1956).

Source: *https://rmets.onlinelibrary.wiley.com/doi/10.1002/qj.49708335521.*

5.8. LARGEST RAINFALL FROM TROPICAL STORMS/HURRICANES

Hurricanes and tropical storms are noted for dumping heavy to extremely heavy rain over a period of days. The storm's converging moisture-laden air is cooled through adiabatic expansion as it rises, and once drenched, it releases its plentiful moisture as rain. When air vapor condenses into moisture, it releases a lot of heat, which is then used to power the tropical storm system. Some storms attack land locations after they leave their formation area. Rainfall over land begins while the storm's center is around 500 kilometers from the shore, according to research. Even after the storm hits the coast and moves inland, heavy rain continues to fall along and near the storm's path. Heavy rainfall linked with tropical storms on a given day can cover an area of up to 400,000 km^2 depending on the paths of tropical storms (Bao, 1987). Tropical storm maximum rainfalls can be shockingly large, ranging from 50 to 80 cm in a single day. The amount of rain that falls varies from storm to storm and location to location. Knowing the highest rainfalls from tropical storms is quite useful. The magnitudes of the world's highest rainfalls are useful not just in the planning of an extensive range of water resources projects, but also in determining the physical top boundary to the amount of rain that may fall over a given area. Table 5.2 shows the highest point and area rainfalls through hot storms across various locations of the world for a one-day period. Table 5.2 compares the maximum known one-day rainfall from hurricanes, hot storms, and typhoons across various sizes of area in numerous locations of the world.

Table 5.2. Principal Areal Rainfalls (cm) for One-Day Duration Related to Hurricanes and Tropical Storms

Country	**Area (km^2)**							**Tropical Storm**
	10000	**5000**	**2000**	**1000**	**500**	**100**	**10**	
China	41	53	66	74	80	90	97	August 7, 1975
India	54	65	77	85	90	97	99	July 2, 1941
USA	46	65	80	85	87	97	98	Sept. 6, 1950
Australia	46	56	66	73	77	85	91	1898

Table 5.3. Largest Rainfall Values Associated with Tropical Storms

Duration	Amount (cm)	Storm Date	Location
72 hours	324 cm	24–27 January, 1980	Grand-Ilet, La Reunion Island
48 hours	247 cm	8–16 April, 1958	Aurere, La Reunion Island
24 hours	183 cm	7–8 January, 1966	Foc-Foc, La Reunion Island
12 hours	115 cm	7–8 January, 1966	Foc-Foc, La Reunion Island
10 days	508 cm	18–27 January, 1980	Commerson, La Reunion Island
5 days	80 cm	6–10 August, 1940	Abbeville, Louisiana

On September 6, 1950, at Yankeetown, Florida, the largest documented Atlantic hurricane one-day point rainfall figure was 98 cm. The greatest typhoon in the Pacific Ocean was 97 cm in Linzhuang, China on August 7, 1975, while the largest in the Bay of Bengal was 99.0 cm in Dharampur, India on July 2, 1941. The world's biggest recorded rainfall levels for various durations linked with tropical storms are listed in Table 5.3.

5.9. CLIMATIC EFFECTS OF HURRICANE ACTIVITY

For the period 1901–1999 in the Atlantic Ocean, the incidence of hot storms and those that reached hurricane intensity is given. During the 99-year span, the year with the most tropical storms, including hurricanes, was 1933, with 21, and the year with the least was 1914, with one. The cause of these fluctuations in the ocean-atmosphere system appears to be non-seasonal. El Nino-Southern Oscillation (ENSO) is a main ocean-atmosphere spectacle that effects weather not only in the immediate vicinity of its incidence but also in distant regions via a complex interacting process. It appears that ENSO is also persuading the changeability of tropical storm production in some oceanic zones. In this perspective, it would be interesting to learn if ENSO influences the frequency of tropical storms in the east Pacific and Atlantic areas of the United States. It is possible to do so by comparing the frequency of hurricanes and hot storms in the Atlantic Ocean through non-ENSO and ENSO years. Before looking into the relationship among ENSO occurrences and hot cyclone frequencies, it is vital to first grasp what ENSO is.

5.10. GLOBAL WARMING AND HURRICANES

Methane, carbon dioxide and other greenhouse gases are expected to warm the climate by 3.6°F (2°C) by 2100, according to the Intergovernmental Panel on Climate Change (IPCC). As a result, the hot sea surface temperature may rise. Because hot storms get their energy from warm tropical waters, a rise in surface ocean temperature could result in frequently and powerful typhoons, tropical cyclones, and hurricanes. Hot cyclone intensity and frequency, on the other hand, are subject to natural climate variability (Gray, 1990). Several researchers have claimed that the recent effects of hurricanes Andrew in 1992 and Hugo in 1989 are proof of global warming. In this context, some researchers wanted to discover if global warming or a warmer world will have an impact on the frequency of tropical storms. Using a Global Climate Model, several researchers explored the association between the occurrence of tropical cyclones and global warming (GCM). A GCM is a climate model that models the interaction of radiation, wind, temperature, and seas to form climate. They are used to forecast the effects of rising greenhouse gas levels. Royer et al. (1998) used a coupled ocean-atmosphere GCM to show that global warming will result in a 10% increase in the number of tropical cyclones in the northern hemisphere and a 5% increase in the southern hemisphere. Druyan et al. (1999) concluded that as the world warms, more tropical cyclones will form across the North Pacific Ocean, the North Atlantic, and east of the Philippines Ocean. Though some modelling studies have suggested that tropical storm frequency may increase as the world warms, the majority of studies have expected that the link among tropical cyclone formation and environmental characteristics will remain static. Henderson-Sellers et al. (1998), for example, recently studied the influence of global warming on hot cyclones and found no significant variations in the area or worldwide location of hot storm genesis under greenhouse situations. They also determined that, in a doubled carbon dioxide environment, the peak strength of hot cyclones may rise by 5% to 10% in wind speeds, but those other factors, like lapse rate and ocean spray changes, will act to decrease changes in wind speeds.

REFERENCES

1. Anthes, R. A., Corell, R. W., Holland, G., Hurrell, J. W., MacCracken, M. C., & Trenberth, K. E., (2006). Hurricanes and global warming—Potential linkages and consequences. *Bulletin of the American Meteorological Society, 87*(5), 623–628.
2. Bao, C., (1987). *Synoptic Meteorology in China* (Vol. 1, pp. 1–23). Springer, Berlin.
3. Camargo, S. J., & Sobel, A. H., (2004). Formation of tropical storms in an atmospheric general circulation model. Tellus A: *Dynamic Meteorology and Oceanography, 56*(1), 56–67.
4. Chen, S. E., Leeman, M. E., English, B. J., Kennedy, A. B., Masters, F. J., Pinelli, J. P., & Natarajan, C., (2016). Basic structure system rating of post-super Typhoon Haiyan structures in Tacloban and East Guiuan, Philippines. *Journal of Performance of Constructed Facilities, 30*(5), 04016033.
5. Druyan, L. M., Lonegran, P., & Eichler, T., (1999). A GCM investigation of global warming impacts relevant to tropical cyclone genesis. *Int. J. of Climatology*, *19*, 607–618.
6. Du, Z., & Selig, M. S., (2000). The effect of rotation on the boundary layer of a wind turbine blade. *Renewable Energy, 20*(2), 167–181.
7. Duke, W. L., (1985). Hummocky cross-stratification, tropical hurricanes, and intense winter storms. *Sedimentology, 32*(2), 167–194.
8. Dunn, G. E., & Miller, B. I., (1964). *Atlantic Hurricanes* (Vol. 1, pp. 377). Louisiana State University Press, Baton Rouge, LA.
9. Elsner, J. B., (2006). Evidence in support of the climate change-Atlantic hurricane hypothesis. *Geophysical Research Letters, 33*(16), 1–20.
10. Emanuel, K. A., (1987). The dependence of hurricane intensity on climate. *Nature*, *326*, pp. 483–485.
11. Friend, D. B., & Abbott, D. C., (1986). The theory of radiatively driven stellar winds. III-Wind models with finite disk correction and rotation. *The Astrophysical Journal, 311*, 701–707.
12. Fussell, E., Curran, S. R., Dunbar, M. D., Babb, M. A., Thompson, L., & Meijer-Irons, J., (2017). Weather-related hazards and population change: A study of hurricanes and tropical storms in the United States, 1980–2012. *The Annals of the American Academy of Political and Social Science, 669*(1), 146–167.

13. Gray, W. M., (1990). Strong association between West African rainfall and US landfall of intense hurricanes. *Science, 249*, pp. 1251–1256.
14. Gray, W. M., Sheaffer, J. D., & Landsea, C. W., (1997). Climate trends associated with multidecadal variability of Atlantic hurricane activity. In: *Hurricanes* (Vol1., pp. 15–53). Springer, Berlin, Heidelberg.
15. Hebert, P. J., Jarrell, J. D., & Mayfield, M., (1997). *The Deadliest, Costliest, and Most Intense United States Hurricanes of This Century* (Vol. 1, pp. 1–24). NOAA, Tech. Memorandum NWS, TPC-1, National Hurricane Center, Florida.
16. Henderson-Sellers, Zhang, A., Berz, H., Emmanual, G., Gray, K. A., Landsea, W., Holland, C., Lighthill, G., Shieh, J., Webster, S. L., & McGuffee, P., (1998). Tropical cyclones and global climate change: A post IPCC assessment. *Bulletin AMS, 79*, pp. 19–38.
17. Jáuregui, E. R. N. E. S. T. O., (2003). Climatology of landfalling hurricanes and tropical storms in Mexico. *Atmósfera, 16*(4), 193–204.
18. Kantha, L., (2006). Time to replace the Saffir-Simpson hurricane scale? *Eos, Transactions American Geophysical Union, 87*(1), 3–6.
19. Khain, A., Cohen, N., Lynn, B., & Pokrovsky, A., (2008). Possible aerosol effects on lightning activity and structure of hurricanes. *Journal of the Atmospheric Sciences, 65*(12), 3652–3677.
20. Khromov, S. P., (1964). *Meteorology and Climatology for Geography Faculties* (Vol. 1, pp. 499). Leningrad.
21. Kinney, D. K., Miller, A. M., Crowley, D. J., Huang, E., & Gerber, E., (2008). Autism prevalence following prenatal exposure to hurricanes and tropical storms in Louisiana. *Journal of Autism and Developmental Disorders, 38*(3), 481–488.
22. Lourensz, R. S., (1981). *Tropical Cyclones in the Australian region-July 1909 to June 1980* (Vol. 1, p. 94). Bureau of Meteorology Australia.
23. Lugo, A. E., (2000). Effects and outcomes of Caribbean hurricanes in a climate change scenario. *Science of the Total Environment, 262*(3), 243–251.
24. McGregor, G. R., (1995). The tropical cyclone hazard over the South China Sea, 19701989: Annual spatial and temporal characteristics. *Applied Geology, 15*, pp. 35–52.
25. Michener, W. K., Blood, E. R., Bildstein, K. L., Brinson, M. M., & Gardner, L. R., (1997). Climate change, hurricanes and tropical storms, and rising sea level in coastal wetlands. *Ecological Applications, 7*(3), 770–801.

26. Mitchell, C. L., (1932). West Indian hurricanes and other tropical cyclones of the North Atlantic Ocean. *Monthly Weather Review, 60*(12), p. 250.
27. Molinari, J., Moore, P., & Idone, V., (1999). Convective structure of hurricanes as revealed by lightning locations. *Monthly Weather Review, 127*(4), 520–534.
28. Molinari, R. L., & Mestas-Nuñez, A. M., (2003). North Atlantic decadal variability and the formation of tropical storms and hurricanes. *Geophysical Research Letters, 30*(10), 1–20.
29. Mousavi, M. E., Irish, J. L., Frey, A. E., Olivera, F., & Edge, B. L., (2011). Global warming and hurricanes: The potential impact of hurricane intensification and sea level rise on coastal flooding. *Climatic Change, 104*(3), 575–597.
30. Neumann, C. J., (1993). *Global Overview in Global Guide to Tropical Cyclone Forecasting* (Vol. 31, pp. 1–23). WMO TC-No. 560, Report No. TCP-31, World Meteorological Organization, Geneva.
31. Nordhaus, W. D., (2010). The economics of hurricanes and implications of global warming. *Climate Change Economics, 1*(01), 1–20.
32. Paerl, H. W., Steppe, T. F., Buchan, K. C., & Potts, M., (2003). Hypersaline cyanobacterial mats as indicators of elevated tropical hurricane activity and associated climate change, AMBIO. *A Journal of the Human Environment, 32*(2), 87–90.
33. Petterssen, S., (1969). *Introduction to Meteorology*. McGraw Hill Book Company, Inc., New York.
34. Quinn, W. H., & Burt, W. V., (1972). Use of Southern Oscillation in weather prediction. *J. Appl. Meteorol.*, *11*, pp. 616–628.
35. Quinn, W. H., Zopf, D. O., Short, K. S., & Kuo, Y. R. T. W., (1978). Historical trends and statistics of the Southern oscillation, El Nino and Indonesian droughts. *Fish. Bull.*, *76*, pp. 663–678.
36. Riehl, H., (1954). *Tropical Meteorology* (Vol. 1, pp. 1–20). McGraw-Hill Book Company, New York.
37. Royer, J. F., Chauvin, B., Timbal, P., & Grimal, D., (1998). A GCM study of the impact of greenhouse gas increase on the frequency of occurrence of tropical cyclones. *Climate Change*, *38*, pp. 307–343.
38. Shepherd, J. M., & Knutson, T., (2007). The current debate on the linkage between global warming and hurricanes. *Geography Compass, 1*(1), 1–24.

39. Simiu, E., Vickery, P., & Kareem, A., (2007). Relation between Saffir-Simpson hurricane scale wind speeds and peak 3-s gust speeds over open terrain. *Journal of Structural Engineering, 133*(7), 1043–1045.
40. Tannerhill, I. R., (1956). *Hurricanes* (Vol. 1, pp. 1–20). Princeton University Press, New Jersey.
41. Tisdale, C. F., & Clapp, P. F., (1963). Origin and paths of hurricanes and tropical storms related to certain physical parameters at the air-sea interface. *Journal of Applied Meteorology and Climatology, 2*(3), 358–367.
42. U. S. Weather Bureau, (1956a). *The Hurricanes* (Revised edition, Vol. 1, pp. 1–27). Washington, D.C.
43. U.S. Weather Bureau, (1956b). *Rainfall Associated with Hurricanes* (Vol. 3, p. 305). National Hurricane Research Project, Report No. 3, U.S. Department of Commerce, Washington, D. C.
44. Walsh, K., & Pittock, A. B., (1998). Potential changes in tropical storms, hurricanes, and extreme rainfall events as a result of climate change. *Climatic Change, 39*(2), 199–213.
45. Xiang, B., Lin, S. J., Zhao, M., Zhang, S., Vecchi, G., Li, T., & Chen, J. H., (2015). Beyond weather time-scale prediction for Hurricane Sandy and Super Typhoon Haiyan in a global climate model. *Monthly Weather Review, 143*(2), 524–535.

Chapter 6

Meteorological Measurements and Instrumentation

CONTENTS

6.1. INTRODUCTION

The suitable prerequisite for frequency of resolution, accuracy, precision, and measurement of observations depends upon their utilization in the required application. The precision in the measured values is referred to their in-time replication, quantity, and space. Accuracy is generally described quantitatively in terms of the standard deviation of outcomes from repetitive judgments in the same circumstances. A measured quantity could be accurate quantitatively but at the same time imprecise, and vice versa (Figure 6.1). It is specifically subjected to the correlations between frequency, resolution, as well as accuracy. The measurement taken by using a rain gauge could be spatially precise, instead, imprecise in quantity and time. However, measurement taken from the radar is imprecise quantitatively, but precise in space and time (Chandrasekar et al., 1990; Bringi and Chandrasekar, 2001).

6.2. POINT MEASUREMENTS OF RAINFALL USING RAIN GAUGES

6.2.1. Types of Rain Gauges

Maximum hydrological and meteorological services utilize non-recording gauges comprising open vessels having vertical sides, namely right cylinders. Different countries utilize different sizes of orifice and height leading to dissimilar measurements. Graduated flask or 'dipstick' is used as a gauge to measure the depth of precipitation; however, recording gauges are frequently utilized in current years. Three kinds of recording gauges are effectively utilized in operation to record data. In the recording gauge named as *weighing type*, combined weight of rainfall receiving can as well as precipitation collected is recorded constantly. The second type is known as the *float type*, in which rainfall passes through a floating chamber comprising a light float. When the level of water increases, it produces vertical movement in the float which is transferred to a pen on a chart. The third type of recording gauge, i.e., *tipping-bucket type*, comprises a container of light metal distributed in two sections, is balanced over the horizontal axis in unstable equilibrium. After the prearranged quantity of rainfall, a relay contact is tipped over by a bucket generating a measurement of the total number of tips produced. Every produced tip might be 0.50 mm, 0.20 mm, or 0.10 mm. Further several kinds of recording gauges could utilize electronic or optical mechanisms to provide a quick response (Norbury, 1974).

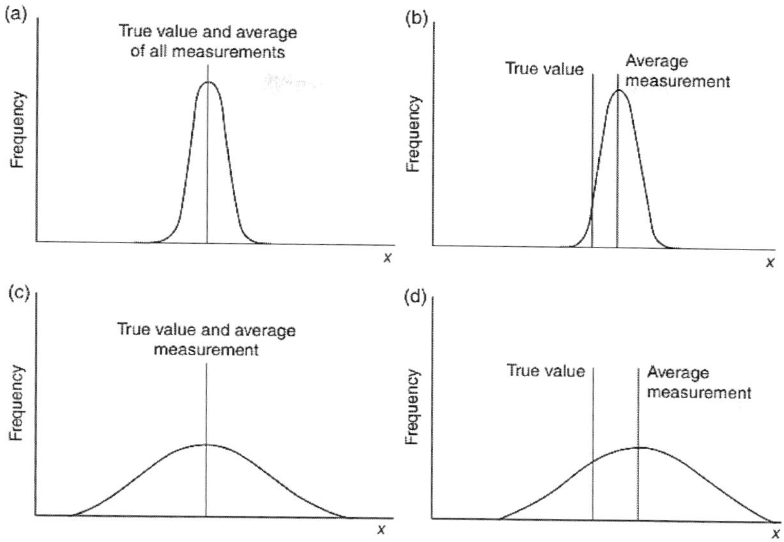

Figure 6.1. Precision and accuracy of different measurements: (a) measurement is accurate as well as exact; (b) measurement is precise but not accurate; (c) measurement is not exact but accurate; (d) measurement is neither precise nor accurate (Chatfield, 1983).

Source: *https://www.routledge.com/Statistics-for-Technology-A-Course-in-App lied-Statistics-Third-Edition/Chatfield/p/book/9780412253409.*

A *disdrometer* is a device employed to calculate the distribution of drop size as well as the velocity associated with falling hydrometeors. In addition, some disdrometers are capable to differentiate between rain, hail, and graupel. The preeminent identified type of such disdrometers as shown in (Figure 6.2) is the Joss-Waldvogel disdrometer (Waldvogel and Joss, 1967), an impact instrument involving the measurement of an impacting raindrop through a relationship concerning the mechanical momentum produced by the impacting drop.

2DVD (2D-video-disdrometer) measurement standard is established by using two high-speed line scan cameras by viewing the particles dropping down via an inlet of 25 × 25 cm^2 of housing (Schonhuber et al., 2007). The intersected optical paths have a common area of about 10 × 10 cm^2, establishing a virtual measurement area. Furthermore, the image grid resolution for raindrops is better than 0.20 mm in both horizontal and vertical directions.

Numerous groups have manufactured a refined optical disdrometer. For instance, University of Iowa, USA has developed an optical disdrometer comprising two laser light beams that are spaced vertically by 1 cm distance. Both beams are laser beams with a wavelength of 670 nm that travel through a collecting and culminating lens, respectively. Both beams later pass via a convex lens placed at a 20 cm distance focusing it on a photodetector. Consequently, a computer identifies the signals and starts storing the voltages at 10 kHz. The characteristics of raindrops, such as, drop size distribution (DSD), shape, velocity, and diameter are then calculated from these voltage measurements. The main advantages of this system are user-friendly, low cost, and no need to calibrate.

Strangeways (2004) indicates that amount of precipitation could not be calculated effectively for numerous specific applications. Moreover, he defines the field testing of design for 3 years regarding Folland's recommendation of an aerodynamic rain collector (1988) as shown in (Figure 6.3).

Figure 6.2. The Joss-Waldvogel disdrometer having measurement surface area of 48 cm^2.

Source: *https://onlinelibrary.wiley.com/doi/book/10.1002/9781118414965.*

Though it displayed an aerodynamic advantage compared to a customary 5-inch gauge, this innovative rain collector showed outsplash during heavy rain. Performance of this new rain collector has been explored in turbulent and real-world conditions instead of restricted and abridged environment produced in the wind tunnel or by the mathematical model as done previously. While performing this activity, videos are recorded by utilizing thread tracers to specify the wind path, providing a novel understanding of the multifaceted wind flow across rain gauges. An innovative design ensued, and the field testing of 02 years has revealed that this novel gauge possesses better aerodynamic and evaporative features along with nominal outsplash, proposing its prospective for enhanced measurement of precipitation.

6.2.2. Measurement of Snow and Hail

To find out how deep the snow is that has just fallen, a graduated ruler or scale is used on the open ground. An average of numerous vertical measurements is taken in places having no nomadic snow. In addition, snow depth could be measured by using a fixed container once snow is leveled without compressing it. Supplementary details have been presented by Collier (2000). Furthermore, the equivalent of water corresponding to a snowfall refers to the magnitude of liquid precipitation enclosed in certain snowfall with following approaches (Sevruk, 1992) as given below:

1. Liquid precipitation weighing or melting.
2. **Rain Gauges:** Snow accumulated in the non-recording rain gauge must be liquefied instantaneously.

Figure 6.3. The Folland-shaped gauge with the collecting bottle (Strangeways, 2004).

Source: *https://onlinelibrary.wiley.com/doi/pdf/10.1002/9781118414965.ch5.*

1. **By Using Radioisotope Gauges:** Gamma rays produced by a radioactive source are employed to determine the water equivalent by measuring the gamma radiation attenuation which is proportionated to the depth of the snow as well as the water equivalent. Two vertical tubes of a similar length separated by a distance of 0.50 to 0.70 m are used as the measuring element. One tube holds the originator, and the other tube contains the detector.
2. **Surveys of Naturally Produced Gamma Radiation:** The attenuation of gamma radiation is the basis for it. produced by the radioactive elements present in the ground caused by the snow.

Snowfall density, averagely, falls with the decline in temperature. The density also decreases with the reducing speed of the wind, and therefore these complications may cause discrepancies in the measurements. However, a reduction in inaccuracies may could be attained by considering the average of numerous measurements taken at each individual point. Number of measures required to certify that the anticipated comparative accurateness of the average values could be accomplished is determined from:

$$N = V_x / V_a \quad (1)$$

N corresponds to the total amount of measurements essential to acquire the specific precision, V_a relates to the allowable standard error squared of the approximation of the average, and V_x represents the variance of errors. This is also applied to rainfall measurements.

Generally, a melting snow cover characteristically covers liquid water in the range of 2% to 5% by weight, though in some cases it covers almost 10% of liquid water. Absorbed solar radiation causes the melting of snow, however, the snow surface reflectivity ranges from 90% for recently fallen snow to about 40% in the case of old snow. Indentations made on hail pads by observers have been used to measure the hail. These pads are typically almost 0.50 m × 0.50 m with thickness of 2.50 cm. Their material is enfolded in aluminum foil. The foil indentations determine the hail size. Generally, a dense net of hail pads is essential to deliver consistent approximations of the hail swath; for instance, 96 hail pads could be dispersed in an area of around 260 km^2.

6.2.3. Measurement Error

The considerable systematic error of 3% to 30% or more determines the usage of can-type rain gauges, principally when visible above the ground level (WMO, 1982). The main systematic error components are presented in Table 6.1, and the corrections model is given as

$$P_k = kP_c = k(P_g + \Delta P_1 + \Delta P_2 + \Delta P_3 + \Delta P_4 - \Delta P_5) + P_r \quad (2)$$

where; P_k represents the modified precipitation amount; P_g corresponds to gauge catch; k is a conversion factor as a result of wind field deformation; precipitation amount accumulated in the gauge is denoted by P_c; ΔP_1 to ΔP_5 are corrections in lieu of different systemic error components rendering to Table 6.1, and P_r are random errors generating from observation and instrument.

Table 6.1. Main Systematic Error Components in Measurement of Precipitation and Their Meteorological and Instrumental Factors (WMO, 1982)

SL. No.	Symbolic Represen tation	Error Component	Magnitude	Instrumental Factors	Meteorological Factors
1.	ΔP_3	Evaporation from the container causes a loss.	2% to 10%	The color, age, isolation of the container, and the orifice area, or the funnel type (rigid or removable)	Precipitation type, saturation shortage, and the speed of wind at the gauge rim level after the end of precipitation till its measurement
2.	ΔP_5	Blowing and drifting snow	1%	Orifice area, the shape, and depth of both gauge collector and rim	Snowstorm intensity and duration, the state of snow cover and speed of the wind.
3.	k	Loss because of wind field distortion above the gauge orifice	2%–10%	Orifice area, the shape, and depth of both gauge collector and rim	Wind speed at the gauge rim in precipitation and the precipitation structure

4.	ΔP_4	Outsplash and in splash	0%–4%	The gauge collector's shape and depth, as well as the type of gauge installation	Rainfall intensity and wind speed
5.	$\Delta P + \Delta P_2$	Losses because of internal walls wetting of the container and collector the when collapsed	10%–50%	The same as above, plus the age of both gauge container and collector and color of material	the drying time of the gauge, and the frequency type, and amount of precipitation, of emptying the container

Generally, *k*, and the corrections ΔP_1 to ΔP_5 for a specific gauge are assessed experimentally by comparing field measurements and conducted laboratory tests. Standard meteorological interpretations at the gauge site or close to it are used to determine the meteorological factors.

Figure 6.4 has been created from site data displaying the variation between the catch of a 5-inch gauge and the Folland-shaped gauge (FSG). During low-intensity rain, a vibrant aerodynamic gain is shown by the FSG over the 5-inch gauge (represented by the upper linear trend line). The alteration in catch rises from 1.80% to about 6.70% with the increase in speed of wind (6 m s^{-1}), an upturn of 4.90%. The trend line does not pass via zero since evaporation is not entirely recompensed.

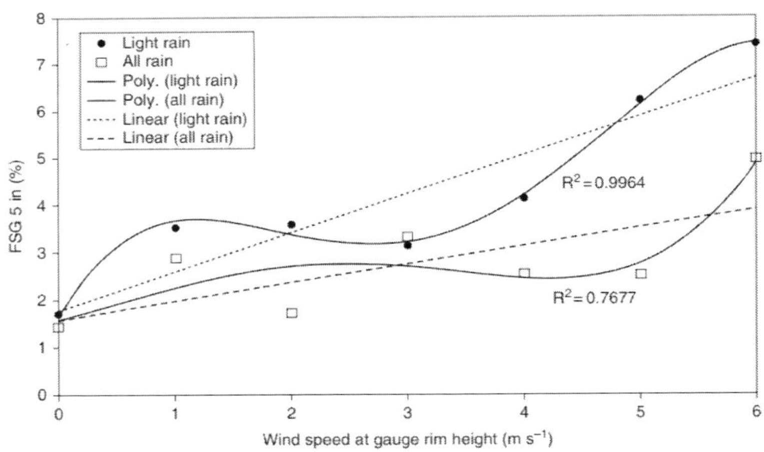

Figure 6.4. The percentage difference in catch for increasing wind speed among the 5-inch gauge and the Folland-shaped gauge (FSG) (Strangeways, 2004).

Source: *https://www.amazon.com/Hydrometry-Principles-Practice-Reginald-H erschy/dp/0471973505.*

It has been stated by Strangeways (2004) that keeping walls low about a gauge to decrease the effects of wind that flows, and the FSG's shallow angled funnel, might cause outsplash during dense rain, and this does befall as validated by Folland (1988). Moreover, in outsplash, raindrops 'bounce' out of the collector as in case of hail. The raindrops disintegrate into several 100 tiny droplets upon impact with a moistened surface. Some droplets leave the collector and some of them fall back or strike the side walls.

6.3. AREAL RAINFALL MEASUREMENT BY RAIN GAUGE

In a study, Shaw (1994) abridges the ways that the measurements taken by rain gauge could be utilized to approximate areal rainfall. The simplest method could be employed in which gauge measurements are added for a certain duration and are then divided by the total amount of gauges used. This is adequate providing the gauges are placed at equal distances and there is no topographical variation in the area. A more refined technique is to use the *Thiessen polygons*. In this technique, individual gauge measurements are initially weighted by the catchment area fractions characterized by each rain gauge. Another more precise technique implicates the use of isohyetals, but a skilled analyst is a prerequisite. Lastly, a multi-quadratic surface resulting from the web of gauges could be used to demonstrate the rainfall surface by employing a computer program.

6.4. RAINFALL MEASUREMENT BY USING RADAR NETWORK

6.4.1. Fundamentals

Weather radar networks, since last 30 years, have been extensively used as a primary part of operational meteorological observing systems to take the qualitative measurements of rainfall for weather forecasts. Nevertheless, operational hydrologists are still reluctant to use these measurements quantitatively. However, this problem is being resolved to a greater extent by improving the algorithm and by using dual-polarization radar. In the current section, evaluation of the recent accuracy attained and different methodologies to radar measurements are emphasized on the basis of Collier (2000) and Illingworth (2004) reviews.

Energy pulses are transmitted by a radar beam rotating around a vertical axis. The backscattered energy from the particles of precipitation is received by the radar in volumes of the size that depends on the radar beamwidth and generated pulse magnitude higher than the ground (Figure 6.5). The total sum of the attenuations from the different scatterers (hydrometeors) at the antenna system of radar is represented by the radar equation (Probert, 1962; Collier, 1996) given below. Considering λ is the radar wavelength, a spherical raindrop has a diameter of D., $\sigma_b(D)$ is backscattering cross-section, and $\sigma_a(D)$ is the overall attenuating cross-section are relative to D^6/λ^4, supposing Rayleigh scattering (small drops). The radar reflectivity factor Z is a physical parameter that is defined as

$$Z = \int_0^{D_{max}} N(D)D^6 dD$$

Where; the DSD within the resolution cell is denoted by N(D) (D in mm, Z in mm^6m^{-3}, $N(D)$ in $mm^{-1}m^{-3}$). The backscattered power received is proportional to Z if there is no attenuation along the radar path and Rayleigh theory is used. Nevertheless, when the ratio $\pi D/\lambda$ turns out to be larger than 0.1 that is for considerably big raindrops and hail, Mie theory is employed. A corresponding radar reflectivity factor Z_e similar to Z for light rain is used to contain this effect.

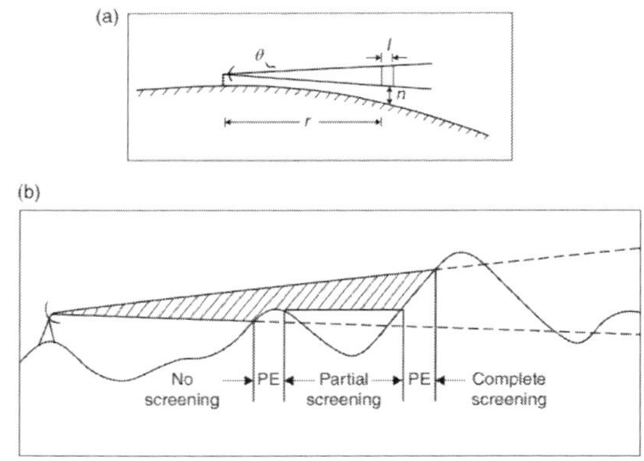

Figure 6.5. (a) The Radar beam geometry in relation to the Earth's curvature. The radar pulse has a length of l and is at a range of r from the radar. The height of the radar beam's base at range r and the angular width of is h. (b) PE (permanent echoes) and screening instigated through hills near radar site (Browning and Collier, 1982, 1989).

Source: *https://agupubs.onlinelibrary.wiley.com/doi/abs/10.1029/RG027i003p00345.*

When the liquid precipitation fills equivalently to the pulse volume, the mean power at range r returning from the precipitation is proportional to Z/r^2, while Z the radar reflectivity factor is represented as the summation of the diameters of precipitation particle outstretched to the sixth power. Z is interrelated to the rainfall rate R and is given as

$$Z = AR^B$$

where; A and B are determined by the rainfall. However, $A = 200$ and $B = 1.6$ are often used. An illustration of reflectivity measurements is presented in Figure 6.6.

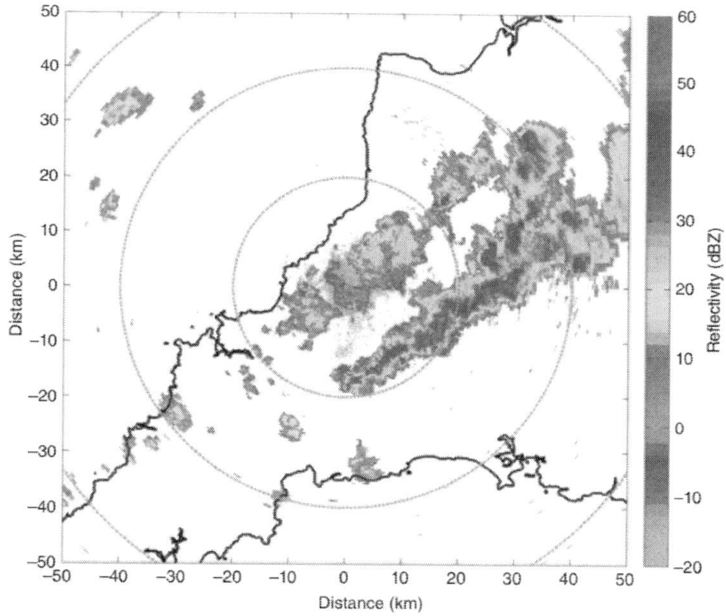

Figure 6.6. The reflectivity (dBZ) measured by the NCAS mobile radar on 3 August 2013 during the convective precipitation experiment (COPE) at a constant elevation angle of 2.5° (Blyth et al., 2015).

Source: *https://rmets.onlinelibrary.wiley.com/doi/full/10.1002/met.1492.*

6.4.2. Radar Measurements Errors

Certain difficulties while taking measurements from radar rise because of the features of the radar and its location. The preservation of a steady system is exceedingly significant. According to Joss and Waldvogel

(1990), the combined inaccuracy of transmitted power, noise figure, antenna characteristics, receiver chain amplification, and analog-to-digital conversion is expected to be better than 2 dB, corresponding to a 36% mistake in precipitation rate.

Usually, for surface precipitation measurements a low horizon (0.5° or less) is essential. Though the significance of this is related to the key beam of radar or its side lobes that could interconnect the ground (Figure 6.5(b)).

This results in screening (occultation) as well as strong insistent ground echoes (ground clutter), triggering radar beam partial loss. The ground clutter map collected during the absence of precipitation could facilitate evade ground clutter and achieve interpolation across it (Figure 6.6). Equally provided around 60% of the beam rests outside screening areas corrections might be applied by it. Still, inconsistent refraction of the radar beam instigated through dry warm air covering cooler, further ground echoes identified as irregular propagation echoes (anaprop) could be originated by moist air. When a radar beam passes through precipitation, the signal is attenuated. Usage of an extended wavelength of 10 cm (S-band) eludes this problem, at 10 cm though the beam width is double in size as at 5 cm (C-band) for the same antenna size, initiating abridged spatial resolution at protracted range. Likewise, precipitation echoes detected at 10 cm distance are 12 dB weaker compared to those witnessed at 5 cm distance for the similar yield power as indicated by Ulaby et al. (1981). A 3 cm radar (X-band) is more operative than a 5 cm radar partaking in similar beam width nevertheless will undergo more attenuation. Nevertheless, the usage of X-band radars has developed more probable because of the improvement of processes utilized for attenuation improvement grounded on polarization parameters' application. In addition, radomes are used to shield the radars from snow, ice, and strong wind. Yet wetting of radomes shielding radars of wavelengths of 5 cm or less could further result in attenuation.

It will be deliberated later, that usage of dual-polarization technology alleviates the influence of attenuation to a high extent making radars of shorter wavelength further attractive.

Besides the glitches in hardware and siting, more complications ascend from the precipitation characteristics. Averaging radar signals above uneven precipitation areas and samples diverge, and substantial disparities in radar pulse volume could initiate errors up to several decibels. Harrison et al., 2009 revealed that the accurateness of data also relies on space and sampling in time, as displayed in Figure 6.7.

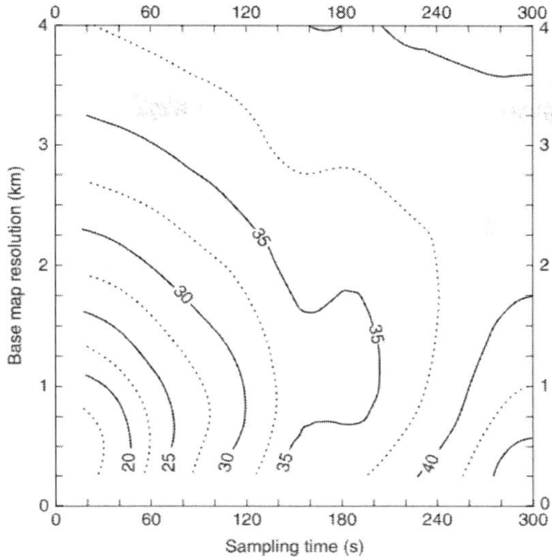

Figure 6.7. A function of the resolution of the reflectivity maps and sample intervals, the absolute inaccuracy (percent) in 5-minute accumulations (Fabry et al., 1994).

Source: *https://www.sciencedirect.com/science/article/abs/pii/0022169494901384.*

The inconsistency of vertical reflectivity may produce huge variations at an increased range between radar measurement of precipitation intensity and the precipitation reaching the ground. Figure 6.8 displays vertical profiles observed by radar in convective and extensive rain at different ranges. Enormous deviancies could result from these profiles on specific occasions, but they designate the precipitation percentage detected by radar permitting for the Earth's curvature. Complications of taking measurements in the bright band (where snow liquefies to form rain) or shallow rain are shown in Figure 6.8(b and c). In addition, orographic growth of rain could also be challenging to measure within 0.5 km of the ground, as presented in Figure 6.8(d).

As stated in Eqn. (4), radar reflectivity observations are connected to rainfall measurements. Battan (1973) lists 69 published relationships that can be found in different types of rainfall. Fortunately, the actual relationships for similar types of rainfall are similar, and Table 6.2 shows typical relationships.

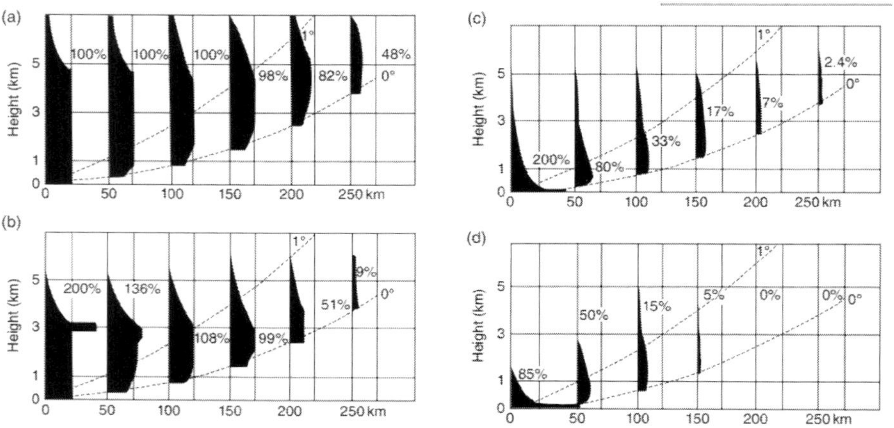

Figure 6.8. Vertical profiles observed by radar in (a) convective rain; (b) widespread rain; (c) snow; (d) orographic rain. Numbers shown represent the rain rate. Assume a radar having 1° beam in a flat country. Much less precipitation is observed at far ranges if the radar is lower than adjacent obstacles. Dashed lines 0° and 1° represent the main section of a 1° beam raised at 0.5° at the top and bottom (Joss and Waldvogel, 1990; Browning and Collier, 1989).

Source: *https://www.scirp.org/(S(351jmbntvnsjt1aadkposzje))/reference/ReferencesPapers.aspx?ReferenceID=1736875.*

Table 6.2. Distinctive Relationships Among Z and R

SL. No.	Equation	Precipitation Type	References
1.	$1780R^{2.21}$	Snowflakes	Sekhon and Srivastava (1970)
2.	$31R^{1.71}$	Orographic rain	Blanchard (1953)
3.	$486R^{1.37}$	Thunderstorm rain	Jones (1956)
4.	$200\ R^{1.6}$	Stratiform rain	Marshall and Palmer (1948)
5.	$2000R^{2.0}$	Aggregate snowflakes	Gunn and Marshall (1958)
6.	$500R^{1.5}$	Thunderstorm rain	Joss et al. (1970)
7.	$140R^{1.5}$	Drizzle	Joss et al. (1970)
8.	$250R^{1.5}$	Widespread rain	Joss et al. (1970)

By considering the work of Zawadzki (1984), Collier (1986), Joss, and Waldvogel (1990), and Rosenfeld et al. (1993), it is evident that the errors in radar measurements of rainfall are not based on Uncertainties in the distribution of drop sizes. Doneaud et al. (1984) established an exceptional correlation between the ATI (area-time integral) of the area surpassing a particular optimal edge reflectivity value and integrated convective rainfall volume.

6.4.3. Adjustment Using Rain Gauges

The utmost frequently employed the technique of regulating the radar assessments of reflectivity relics the use of measurements independently taken from rain gauges. Several diverse techniques are employed, from alteration factors resulting from simple applications of the ratios of rain gauge to radar values (Collier et al., 1983) to the further intricate techniques proposed by Doneaud et al. (1984).

A method that has been employed effectively involves matching the unconditional prospects of the intensity of rainfall acquired from radar reflectivity and rain gauges (Rosenfeld et al., 1993, 1994) and is called the PMM (probability matching method). It has been revealed by Rosenfeld et al. (1994) that 600 mm of collected rainfall was obligatory to provide a suitable empirical demonstration of the rain gauge rainfall likelihood distributions and the radar. This renders to almost 600 samples subjected to the time interval employed. Nevertheless, Seed et al. (1996) has validated that when only 500 samples are considered, the PMM sorts systematic underestimates related to elevated high rainfall rates. This partiality only declines deliberately by increasing the number of samples (Liu and Chandrasekar, 2000; Kim et al., 2010).

The PMM includes converting the data to dBZ received from radar and rain gauge utilizing the general formula.

$dBZ = 10\log_{10}(fV^{p})$ where; V denotes the amount of rain that will fall (mm h^{-1}); *f* is a factor having a characteristic value of 200; and *p* is a power having a value of 1.6.

In the latest applications averaging intervals of 5 minutes to 1 hour are used. For instance, 9 minutes have been used by Rosenfeld et al. (1994) whereas 30 minutes have been used by Atencia et al. (2010). The scattering between the radar data (real data) and rain gauge data (point data) presents sampling errors as stated by (Villarini et al., 2008; Kitchen and Blackall, 1992). Hourly rainfall accurateness wholes over river sub-catchments-full curves which are attained by using synchronized radar positioned inside 50 km of a site of interest-as presented in Figure 6.9. The accuracy attained with networks of rain gauges in the nonappearance of radar is shown by dotted curves. The accuracy is drawn as a function of rain gauge density for both sets of curves. Rain gauge networks' accuracy is influenced significantly by the type of rain, as revealed by the major deviations among the four dotted curves. Conversely, the radar measurements are approximately self-determining of the type of rain (Geotis, 1963; Collier and Larke, 1978).

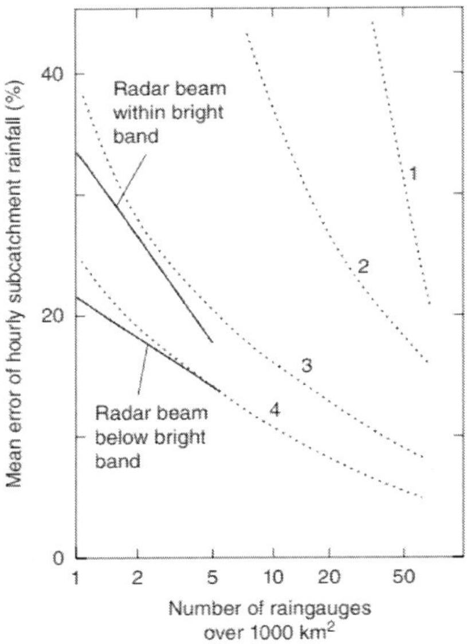

Figure 6.9. The mean error of hourly rainfall totals in an average area 60 km^2 of river catchments determined in different kinds of rainfall conditions from radar measurements, plotted against the number density of rain gauge sites (solid lines).

Note: Also shows a comparison of the mean error of the hourly sub-catchment totals as determined by network of rain gauges without using radar, plotted against the number density of rain gauge sites (dotted lines). The measurement errors for the rain gauge network are represented by four dotted curves in the existence of (1) exceedingly rare showers; (2) common showers; (3) common spreading rain; and (4) exceptionally uniform rain. For all curves, the mean error is defined as the mean value of the difference between the estimated rainfall and the 'best estimate,' regardless of sign (Collier, 1977; Browning, 1978, 1983).

Source: *https://onlinelibrary.wiley.com/doi/pdf/10.1002/9781118414965.ch5.*

6.5. MEASUREMENT OF SOIL MOISTURE

6.5.1. Methodologies

Soil moisture is measured using two methodologies (Hardaker and Auer, 1994):

- The ratio between the mass of water present in the soil sample and the dry mass or volume of the soil sample is known as soil water content. The dry bulk density of the soil sample is a coefficient that connects these two concepts.
- The potential energy of the water content in the soil sample, which includes gravitational pressure and osmotic components, is known as soil water potential. In most cases, the soil-water mixture is considered to be locally homogeneous, and the osmotic potential is ignored (Meischner et al., 2004; Matrosov, 2010).

Due to high spatial variability, the soil moisture must be determined at several test points of an area under analysis. A brief review of different techniques is presented in WMO (1994) and Herschy (1999).

6.5.2. Measurement by Gravimetric Method

The gravimetric method is employed to measure the soil moisture directly. It measures the moisture content by weighing the soil sample after and before drying. The sample of soil is deliberated to be completely dry at 105°C while the weight stays the same. Numerous types of sampling equipment could be used to measure the moisture content.

For instance, the simplest equipment is named a hand auger an aluminum pipe used for sampling depths as deep as 17 meters. Instead, a core barrel, drive sampler, or a soil sampling tube can be employed, which delivers a benefit in a sampling of soil by acquiring volumetric samples for computing moisture content by volume (Illingworth and Blackman, 2002; Meischner et al., 2004).

The open-drive sampler comprises a core barrel having 50 millimeters inner diameter and 100 millimeters length, along with extension tubes of 25 millimeters in diameter and 1.5 meters of length to carry out sampling at depth. 50 millimeters long brass cylinder liners are used to preserve the unobstructed core samples. Later the samples are detached by pushing a plunger of the core barrel (Pollock and Wilson, 1972).

6.5.3. Measurement by Electrical Resistance Method

Soil moisture content is also measured by finding the electrical resistance of a porous material block in a state of equilibrium with the soil in terms of moisture. Two electrodes are firmly attached to the block of material like nylon, plaster of Paris, or fiberglass. The resistance of contact remains the same, and once the block is sited in the soil and acquires moisture equilibrium, it modifies to promote variations in soil moisture after a momentary lag time. Generally, soil moisture blocks are deliberated of utmost reliability in the range of low moisture amount. Their appropriateness for calculating moisture of a sample is constrained by hysteretic effects, as that calibration is dependent on the temperature and density of the soil (Jatila, 1973; Ryzhkov and Zrnic, 1995).

6.5.4. Measurement by Neutron Method

The neutron approach determines how much water is present per unit of soil volume. This technique involves the measurement of the slowed-down neutrons by a fast source of neutron (Greacen, 1981) emitted into the soil sample. The loss of energy is considerably large when neutrons collide with atoms of low atomic weight and the loss of energy is related to the number of low atomic weight atoms existing in the sample. The purpose of these collisions of neutrons with the atoms is to convert a fast neutron to a slow-speed neutron. Hydrogen being a primary element present in the soil is principally the part of molecules of water present in the soil. A counter tube senses the number of slow neutrons after the emission of fast neutrons from a tube of radioactive source is electronically specified on a scale. This approach yields a bulb-shaped volume of soil with a radius of 1 to 4 meters, which is proportional to the moisture content of the soil and the activity of the source (Sachidananda and Zrnic, 1986).

6.5.5. Measurement by Gamma-Ray Attenuation Method

The gamma-ray intensity endures an exponential drop when it is emitted into a segment of soil that primarily relies on the water contained in the soil sample, the apparent soil density, and the constants coefficients of attenuation of the soil water. This method involves simultaneously lowering a gamma-ray source, such as cesium-137, and a gamma-ray detector (scintillator photomultiplier) in the soil, as well as a pair of parallel access tubes. The signal can be converted into the apparent wet density of the soil at each measurement level; or, if there is known dry bulk density, the signal could be translated into an extent of the volumetric moisture content of the soil. This is not particular to water only (Seliga and Bringi, 1976; Smith, 1984).

6.5.6. COSMOS–UK Method

At the hectometer horizontal scale Zreda et al. (2008, 2012) explained the cosmic-ray method to identify the area-averaged soil moisture content. A cosmic ray soil moisture probe that remains stationary is used to sense the neutrons produced through cosmic rays in air, soil, and other materials, which are slowed down by principally low atomic weight hydrogen atoms situated mainly in soil and water.

NERC (Natural Environment Research Council), UK, is facilitating the CEH (Center for Ecology and Hydrology) to arrange the COSMOS-UK (Cosmic-ray Soil Moisture Observing System UK) that will provide real-time weather monitoring and field-scale calculations of soil moisture across the UK. At CEH Wallingford the measuring principle is analogous to the neutron probe, however, the COSMOS system depends on naturally taking place neutrons, instead of generated by an artificial source (Stephenson, 1965; Smyth et al., 1999).

All stations of COSMOS-UK are equipped with advanced meteorological and soil monitoring sensors. COSMOS probes developed at COSMOS-UK deliver soil moisture over a vast area content (at a spatial scale of about 700 m). TDT (two near-surface time-domain transmissometry) probes are also used to provide point measurements of moisture content and at three depths from a profiled probe. Near-surface TDT probes are used to determine the soil temperature by a soil temperature profile at five depths. Soil heat flux is offered by two soil heat flux plates. Net radiation is measured from the four separately noted components (entering and departing short and longwave radiation).

6.5.7. Measurement by Dielectric Methods

The soil is a multifaceted mixture of different particles with varying chemical compositions in a matrix of water and air. The dielectric constant (permittivity) of a volume of soil differs from the soil moisture content. Dielectric constant value for soil particles is from 2 to 7, for air is 1, and for water is 80. The corresponding dielectric constant allows the conversion of dielectric values to volumetric soil moisture content when the dry bulk density of soil becomes constant. Straka et al., 2000 stated two methods for estimating the moisture content of the soil which are given below.

TDR (Time-domain reflectometry) calculates the speed of a microwave pulse traveling among a pair of waveguides that are placed in the soil. The speed of pulse depends on the outward dielectric permittivity of the soil among the waveguides (Topp et al., 1980). An increase in the pulse speed indicates the decline in the permittivity, showing a reduction in the soil moisture content (Figure 6.10) (Ulbrich, 1983; Testud et al., 2000).

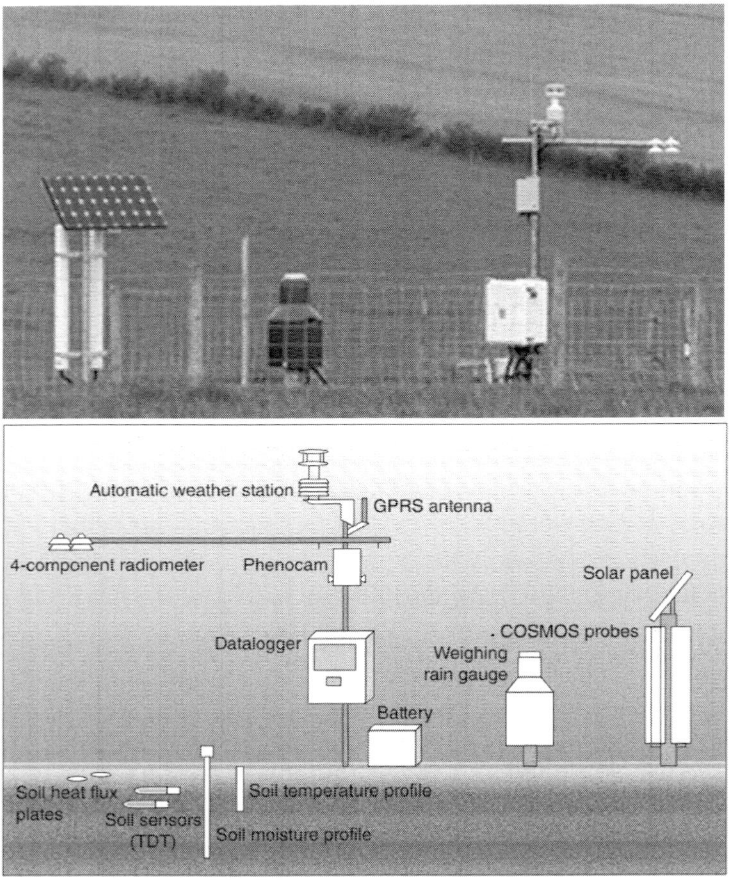

Figure 6.10. Schematic of COSMOS station (courtesy CEH).

Source: https://onlinelibrary.wiley.com/doi/book/10.1002/9781118414965.

The *capacitance method* involves a sensor consisting of electrodes placed in the soil, called electronic oscillator, accompanied by a frequency meter and connecting cables. A capacitor is formed by the electrodes and their nearby soil with a capacitance depending upon the dielectric constant of the soil, ultimately providing a measurement of the soil moisture content. By using a suitable calibration, the fluctuating capacitance is calculated by variations in the frequency of the signal across the capacitor (Waldvogel et al., 1979; Burt, 2016).

6.5.8. Measurement by Tensiometric Method

A tensiometer entails a cup, porous point, or alike attached to a pressure calculating device through a tube. Water is filled in the system. The filled water approaches an equilibrium state with the moisture contained in the adjacent soil. As the soil dries, water runs out of the point, and generates enhanced tension, or pours back inside the point when the soil gets wet and has low tension. These deviations in tension are detected on a measuring device. Various tensiometers could be positioned at numerous depths to permit the assessment of moisture content (Harrison et al., 2016; Portas et al., 2016).

6.5.9. Measurement by Satellite Remote Sensing

Satellite remote sensing is the solitary way of acquiring data that deliver widespread area united information. Satellite-borne radar also provides measurements along with intermediate areas (Gray and Harrison, 2012, 2016; Barnard et al., 2016).

6.6. MEASUREMENT OF EVAPORATION AND EVAPOTRANSPIRATION (ET)

Lysimeters and evaporation pans are employed in networks to calculate evaporation. Several different sorts of evaporation pans are used with circular cross-sections or with square cross-sections. Regular evaporation is calculated as the water level variation among two measurements amended for any precipitation (Nicoll et al., 2012; Aplin et al., 2016).

Soil evaporimeters and lysimeters are used to assess evapotranspiration (ET) through heat-budget - or water approaches, by using the turbulent-diffusing technique, or through applying different empirical formulae (WMO, 1976). Variation among measured ET and contemporaneously measured evaporation from the soil gives the measurement of transpiration of vegetation.

It is not likely to determine ET in natural conditions directly from a vast area. Over an extended time, frame, a water-budget approach could be utilized from a basin having all items of inflow and outflow, excluding ET, can be calculated (WMO, 1985). Instead, aerodynamic, and energy-budget and pan evaporation methods could be employed (WMO, 1966).

6.7. MEASUREMENT OF FLOW: BASIC HYDROMETRY

For flow in pipes and open channels, the discharge equation is:

$$Q = VA$$

where; Q stands for discharge (m^3s^{-1}); V for velocity ($m\ s^{-1}$); and A for the wetted area (m^2). The technique implicates approximating velocity and area directly or indirectly and the Figure 6.11 describes the principle and theory of the weir equation of discharge. The majority of streamflow stations utilize the current meter to determine velocity and the area is directly calculated through a line or sounding rod or other methods. In the case of rivers with a steady bed, the area is related to a function of the stage, though at ultrasonic stations velocity and area are also calculated directly. Velocity and area could be indirectly measured by electromagnetic stations by using the following relation

$$V = \sqrt{2gH}$$

where; H denotes the total head over the barrier in meters. Procedures comprise the sluice gates, Pitot tube, flow in pipes, and open channel orifices. Doppler systems calculate flows directly (Clark, 2016; de Podesta et al., 2018).

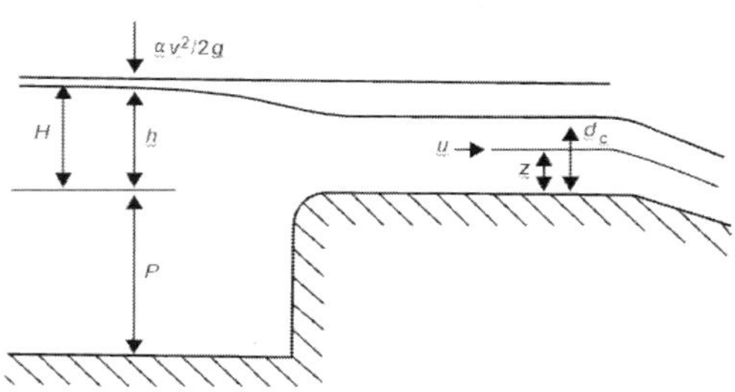

Figure 6.11. The principle and theory of the weir equation of discharge are depicted in the diagram (Herschy, 1999).

Source: *https://www.wiley.com/en-ao/Hydrometry%3A+Principles+and+Practice%2C+2nd+Edition-p-9780471973508.*

6.8. MEASUREMENT OF STREAM DISCHARGE

As revealed in Figure 6.12, the river discharge is calculated by dividing the river cross-section with spaced verticals. The verticals are moved apart to confirm a satisfactory sample of both bed profile and velocity distribution. The discharge is determined from the sum of the products of depth, velocity, and space between the verticals. The total of the products of depth, velocity, and space between the verticals determines the discharge. A number of verticals and spacing are acute for exact measuring. Herschy (1995) recommends 3–4 verticals for a channel width of 0.0–0.50 m, and greater than 20 for a channel width >10 m (Scott et al., 2016).

6.8.1. Use of Stage-Discharge Curve

Precaution must be taken in inferring stage-discharge curves. The most adequate process referred to as the stage-area-velocity technique is proposed by Herschy (1995, 1999) that implies the use of discharge, mean velocity curves plotted, and area against stage on the same graph and it depicted in Figure 6.12.

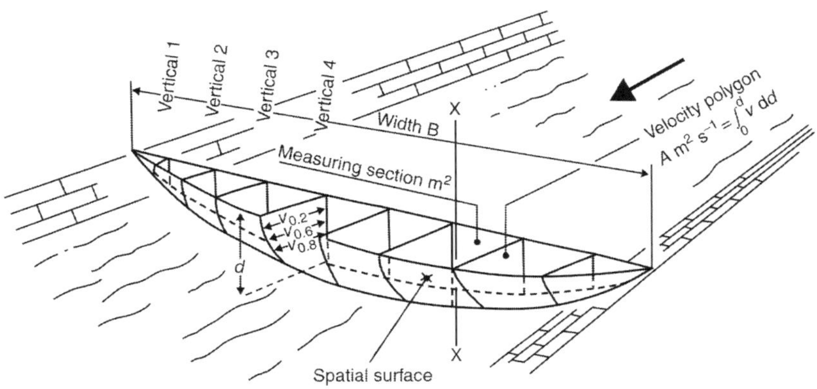

Figure 6.12. The schematic view of calculating cross-section divided into segments across the channel. The sum of the products of depth, velocity, and distance among the verticals provides the discharge (Herschy, 1999).

Source: https://www.wiley.com/en-ao/Hydrometry%3A+Principles+and+Practice%2C+2nd+Edition-p-9780471973508.

This technique relies on the rate of growth of mean velocity at the developed stages declining promptly in such a way that the stage/mean-velocity curve could be stretched with adequate error. The product of mean velocity and area provides the extrapolated discharges as displayed in Figure 6.13 (Harrison, 2011; Marlton et al., 2016).

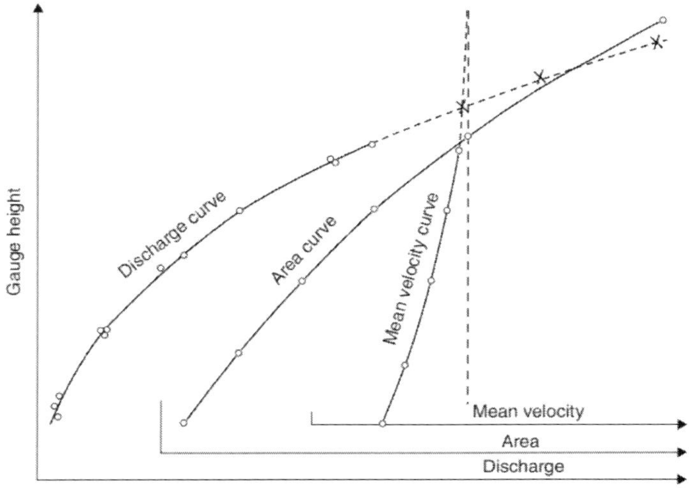

Figure 6.13. Extrapolating a stage-discharge curve using the stage-area-velocity approach (Herschy, 1999).

Source: https://onlinelibrary.wiley.com/doi/pdf/10.1002/9781118414965.ch5.

6.8.2. Measurement by Automated Moving Boat Methods

In this method, a current meter of propeller-type is hanged from a boat at around 1 m beneath the surface, and the boat passes the channel perpendicular to the streamflow as stated by (Smoot and Novak, 1969). During the movement of the boat, the geometry of the cross-section is recorded by an echo sounder and the current meter determines the velocities of boat and stream. The velocity measurement recorded in the cross-section at every sampling vertical is referred to as a vector quantity and is relative velocity previous the meter (Herschy, 1999; Papoutsidakis et al., 2017).

Velocity is also calculated by the broadband acoustic Doppler current profiler by determining the Doppler shift among acoustic signals received off residue particles in the flow. The direction and magnitude of the velocity as well as the depth of the flow are calculated in a water column. The system is typically installed on a boat that measures its comparative location when it passes through the channel (Aplin and Harrison, 2003; Cetinkaya, 2009).

6.9. MODERN TELEMETRY APPROACHES

6.9.1. Use of Telemetry Links

Special data webs are currently accessible to perform digital data transmission. Dial-up internet and computer links are delivered by several companies to provide dedicated links or access on a dial-up basis to an operator. A drawback of telephone lines is their expense, mainly if the links are needed for constant access to data, as in case of flood or bad weather cautions. Currently, telephone lines and cable TV based on fiber optics deliver ready access to the World Wide Web (Collins et al., 1989; Long et al., 2011).

6.9.2. UHF and VHF Radio Links

Wherever telephone lines are not existing, UHF (300 MHz to 3 GHz, 10 cm wavelength) or VHF (30–300 MHz) radio links deliver alternate communication means. Though, authorization for using such frequencies must be acquired from the authorized regulatory body to evade intrusion. Normally UHF frequencies are extensively utilized as these entail a greater bandwidth thereby providing more channels. Before the existence of telephone lines, HF radio (short wavelength) has been extensively used. It transmits by continual radio wave reflections between the ionosphere and the ground covering vast distances. Nevertheless, the technique could undergo dwindling because of weather interfering and sunspot activity. As a result, satellite communication links have mostly substituted this technique (Schnädelbach et al., 2008; Young, 2018). The UHF and VHF radio communication links is graphically explained by Figure 6.14.

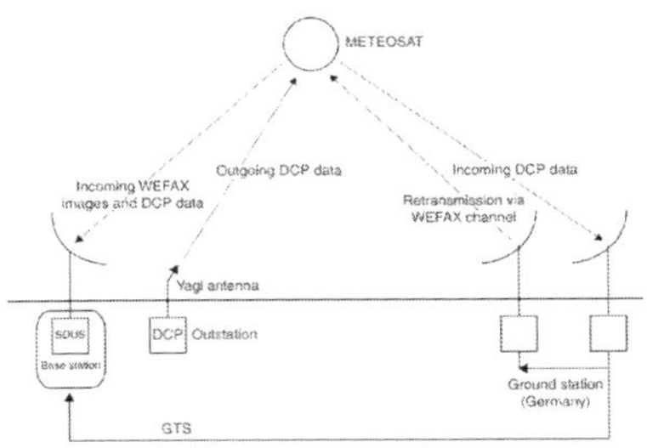

Figure 6.14. The pattern of communication between outstation, METEOSAT, the user and the ground station communicate via the WEFAX retransmission route via the WMO GTS network (Strangeways, 1999).

Source: *https://www.nhbs.com/hydrometry-principles-and-practice-book.*

6.9.3. Use of Satellite Links

Numerous geostationary satellites like the GEOS, GOMS, METEOSAT, and INSAT now can acquire remotely sensed images and transmit the data to the associated ground stations. Similarly, the commercially operated satellites spread out globally, particularly Inmarsat, also deliver data transmission services to various users. The DCP (data collection platform) system delivered by METEOSAT is shown in Figure 6.14. A METEOSAT outstation transmitter functions at a frequency of 402.01 MHz to 402.20 MHz and is focused on the satellite through a Yagi antenna (Hassanien et al., 2020).

REFERENCES

1. Aplin, K. L., & Harrison, R. G., (2003). Meteorological effects of the eclipse of 11 August 1999 in cloudy and clear conditions. *Proceedings of the Royal Society of London; Series A: Mathematical, Physical and Engineering Sciences, 459*(2030), 353–371.
2. Aplin, K. L., Scott, C. J., & Gray, S. L., (2016). Atmospheric changes from solar eclipses. Philosophical Transactions of the Royal Society A. *Mathematical, Physical and Engineering Sciences, 374*(2077), 20150217.
3. Atencia, A., Llasat, M. C., Garrote, L., & Mediero, L., (2010). Effect of radar rainfall time resolution on the predictive capability of a distributed hydrological model. *Hydrol. Earth Syst. Sci. Discuss., 1*(7), 7995–8043.
4. Barnard, L., Portas, A. M., Gray, S. L., & Harrison, R. G., (2016). The National Eclipse Weather Experiment: an assessment of citizen scientist weather observations. *Philosophical Transactions of the Royal Society A: Mathematical, Physical and Engineering Sciences, 374*(2077), 20150220.
5. Battan, L. J., (1973). *Radar Observation of the Atmosphere* (Vol. 1, pp. 1–20). University of Chicago Press, Chicago.
6. Blanchard, D. C., (1953). Raindrop size distribution in Hawaiian rains. *J. Meteorol., 10*, 457–473.
7. Blyth, A. M., Bennett, L. J., & Collier, C. G., (2015). High-resolution observations of precipitation from cumulonimbus clouds. *Meteorol. Appl., 22*, 75–89.
8. Bringi, V. N., & Chandrasekar, V., (2001). *Polarimetric Doppler Weather Radar* (Vol. 1, pp. 1–20). Cambridge University Press, Cambridge.
9. Browning, K. A., & Collier, C. G., (1982). An integrated radar-satellite nowcasting system in the UK. Chapter 1.5. In: Browning, K. A., (ed.), *Nowcasting* (Vol. 1, pp. 47–61). Academic Press, London.
10. Browning, K. A., & Collier, C. G., (1989). Nowcasting of precipitation systems. *Rev. Geophys., 27*, 345–370.
11. Browning, K. A., (1978). Meteorological applications of radar. *Rep. Prog. Phys., 41*, 761–806.
12. Browning, K. A., (1983). Air motion and precipitation growth in a major snowstorm. *Q. J. R. Meteorol. Soc., 109*, 225–242.

13. Burt, S., (2016). Meteorological responses in the atmospheric boundary layer over southern England to the deep partial eclipse of 20 March 2015. *Philosophical Transactions of the Royal Society A: Mathematical, Physical and Engineering Sciences, 374*(2077), 20150214.
14. Cetinkaya, E. K., (2009). Aeronautical gateways: Supporting TCP/IP-based devices and applications over modern telemetry networks. *International Foundation for Telemetering* (Vol. 1, pp. 1–23).
15. Chandrasekar, V., Bringi, V. N., Balakrishnan, N., & Zrnic, D. S., (1990). Error structure of multiparameter radar and surface measurement of rainfall. III: Specific differential phase. *J. Atmos. Oceanic Technol., 7*, 621–629.
16. Chatfield, C., (1983). *Statistics for Technology* (3rd edn., Vol. 1, pp. 1–25). Chapman and Hall, London.
17. Clark, P. A., (2016). Numerical simulations of the impact of the 20 March 2015 eclipse on UK weather. *Philosophical Transactions of the Royal Society A: Mathematical, Physical and Engineering Sciences, 374*(2077), 20150218.
18. Collier, C. G., & Larke, P. R., (1978). A case study of the measurement of snowfall by radar. *Q. J. R. Meteorol. Soc., 104*, 615–621.
19. Collier, C. G., (1977). Radar measurements of precipitation. *Proceedings of the WMO Technical Conference on Instrumentation and Methods of Observation* (Vol. 480, pp. 202–207). Hamburg, WMO Technical Document; World Meteorological Organization, Geneva.
20. Collier, C. G., (1986). Accuracy of rainfall estimates by radar. I: Calibration by telemetering rain gauges. *J. Hydrol., 83*, 207–223.
21. Collier, C. G., (1996). *Applications of Weather Radar Systems: A Guide to Uses of Radar Data in Meteorology and Hydrology* (Vol. 2, pp. 1–35). Wiley-Praxis, Chichester.
22. Collier, C. G., (2000). *Precipitation Estimation and Forecasting* (Vol 467, pp. 1–20). WMO Operational Hydrology Report 467, WMO 887. World Meteorological Organization, Geneva.
23. Collier, C. G., Larke, P. R., & May, B. R., (1983). A weather radar correction procedure for real-time estimation of surface rainfall. *Q. J. R. Meteorol. Soc., 109*, 589–608.
24. Collins, A., Dominy, C., & Chesney, J. R., (1989). Transportable telemetry workstation for future distributed system environments. *International Foundation for Telemetering, 1*, 1–20.

25. De Podesta, M., Bell, S., & Underwood, R., (2018). Air temperature sensors: Dependence of radiative errors on sensor diameter in precision metrology and meteorology. *Metrologia, 55*(2), 229.
26. Doneaud, A. A., Nisov, S. I., Priegrutz, D. L., & Smith, P. L., (1984). The area-time integral as an indicator for convective rain volumes. *J. Climate Appl. Met.*, *23*, 555–561.
27. Fabry, F., Bellon, A., Duncan, M. R., & Austin, G. L., (1994). High resolution rainfall measurements by radar for very small basins: the sampling problem re-examined. *J. Hydrol.*, *161*, 415–428.
28. Folland, C. K., (1988). Numerical models of the rain gauge exposure problem, field experiments and an improved collector design. *Q. J. R. Meteorol. Soc.*, *114*, 1485–1516.
29. Geotis, S. G., (1963). Some radar measurements of hailstorms. *J. Appl. Meteorol.*, *2*, 270–275.
30. Gray, S. L., & Harrison, R. G., (2012). Diagnosing eclipse-induced wind changes. *Proceedings of the Royal Society A: Mathematical, Physical and Engineering Sciences, 468*(2143), 1839–1850.
31. Gray, S. L., & Harrison, R. G., (2016). Eclipse-induced wind changes over the British Isles on the 20 March 2015. *Philosophical Transactions of the Royal Society A: Mathematical, Physical and Engineering Sciences, 374*(2077), 20150224.
32. Greacen, E. L., (1981). *Soil Water Assessment by Neutron Method* (Vol. 1, pp. 1–20). CSIRO Special Publication. CSIRO, Melbourne.
33. Gunn, K., & Marshall, J., (1958). The distribution with size of aggregate snowflakes. *J. Meteorol.*, *15*, 452–461.
34. Hardaker, P. J., & Auer, A. H., (1994). The separation of rain and hail using single polarization radar echoes and IR cloud-top temperatures. *Meteorol. Appl.*, *1*, 201–204.
35. Harrison, D. L., Scovell, A. W., & Kitchen, M., (2009). High-resolution precipitation estimates for hydrological users. *Proc. Inst. Civil Eng. Water Manag.*, *162*, 125–135.
36. Harrison, R. G., (2011). Lag-time effects on a naturally ventilated large thermometer screen. *Quarterly Journal of the Royal Meteorological Society, 137*(655), 402–408.
37. Harrison, R. G., Marlton, G. J., Williams, P. D., & Nicoll, K. A., (2016). Coordinated weather balloon solar radiation measurements during a solar eclipse. *Philosophical Transactions of the Royal Society*

A: Mathematical, Physical and Engineering Sciences, 374(2077), 20150221.

38. Hassanien, A. E., Darwish, A., & Abdelghafar, S., (2020). Machine learning in telemetry data mining of space mission: Basics, challenging and future directions. *Artificial Intelligence Review, 53*(5), 3201–3230.
39. Herschy, R. W., (1995). *Streamflow Measurement* (Vol. 2, pp. 1–26). Chapman and Hall, London.
40. Herschy, R. W., (1999). *Hydrometry: Principles and Practices* (Vol. 2, pp. 1–27). Wiley, Chichester.
41. Illingworth, A., & Blackman, T. M., (2002). The need to represent raindrop size spectra as normalized gamma distributions for the interpretation of polarization radar observations. *J. Appl. Meteorol., 41*, 1578–1583.
42. Illingworth, A., (2004). Improved precipitation rates and data quality by using polarimetric measurements: Chapter 5. In: Meischner, P., (ed.), *Weather Radar Principles and Advanced Applications* (Vol. 5, pp. 130–166). Springer, Berlin.
43. Jatila, E., (1973). Experimental study of the measurement of snowfall by radar. *Geophysica, 12*, 1–10.
44. Jones, D. M. A., (1956). *Rainfall Drop-Size Distribution and Radar Reflectivity* (Vol. 6, pp. 1–25). Research Report 6, Urbana Meteorology Laboratory, Illinois State Water Survey.
45. Joss, J., & Waldvogel, A., (1967). Ein Spektograph fur Niederschlagstrophen mit automatischer Auswertung. *Pure Appl. Geophys., 68*, 240–246.
46. Joss, J., & Waldvogel, A., (1990). Precipitation measurement and hydrology: A review. Battan Memorial and Radar Conference. In: Atlas, D., (ed.), *Radar in Meteorology* (Vol. 1, pp. 577– 606). Chapter 29a. American Meteorological Society, Boston.
47. Joss, J., Schran, K., Thams, J. C., & Waldvogel, A., (1970). *On the Quantitative Determination of Precipitation by Radar* (Vol. 1, pp. 1–28), Zurich.
48. Kim, D.-S., Maki, M., & Lee, D. I., (2010). Retrieval of three-dimensional raindrop size distribution using X-band polarimetric radar data. *J. Atmos. Oceanic Technol., 27*, 1265–1285.
49. Kitchen, M., & Blackall, R. M., (1992). Representativeness errors in comparisons between radar and gauge measurements of rainfall. *J.*

Hydrol., *134*, 13–33.

50. Liu, H. P., & Chandrasekar, V., (2000). Classification of hydrometeors based on polarimetric radar measurements: Development of fuzzy logic and neuro-fuzzy systems, and *in situ* verification. *J. Atmos. Oceanic Technol.*, *17*, 140–164.
51. Long, S. A., Reiger, P. A., Elliott, M. W., Edney, S. L., Knabe, F., & Bernhard, D., (2011). *Application of a Miniature Telemetry System in a Small Gas Turbine Engine* (Vol. 54631, pp. 261–270). In Turbo Expo: Power for Land, Sea, and Air.
52. Marlton, G. J., Williams, P. D., & Nicoll, K. A., (2016). On the detection and attribution of gravity waves generated by the 20 March 2015 solar eclipse. *Philosophical Transactions of the Royal Society A: Mathematical, Physical and Engineering Sciences, 374*(2077), 20150222.
53. Marshall, J. S., & Palmer, W., (1948). The distribution of raindrops with size. *J. Meteorol.*, *5*, 165–166.
54. Matrosov, S. Y., (2010). Evaluating polarimetric X-band radar rainfall estimators during HMT. *J. Atmos. Oceanic Technol.*, *27*, 122–134.
55. Meischner, P., Dotzek, N., Hagen, M., & Holler, H., (2004). Understanding severe weather systems using Doppler and polarization radar. Chapter 6. In: Meischner, P., (ed.), *Weather Radar: Principles and Advanced Applications* (Vol. 1, pp. 167–198). Springer, Berlin.
56. Nicoll, K. A., & Harrison, R. G., (2012). Balloon-borne disposable radiometer for cloud detection. *Review of Scientific Instruments, 83*(2), 025111.
57. Norbury, J. R., (1974). A rapid-response rain gauge for microwave attenuation studies. *J. Technol. Atmos.*, *8*, 245–251.
58. Papoutsidakis, M., Symeonaki, E., Piromalis, D., & Tseles, D., (2017). Modern shipping navigation based on telemetry and communication systems. *International Journal of Computer Applications, 975*, 8887.
59. Pollock, D. M., & Wilson, J. W., (1972). Basin precipitation: Land and lake. *IFYGL Technical Plan, 1*, pp. 107–112.
60. Portas, A. M., Barnard, L., Scott, C., & Harrison, R. G., (2016). The national eclipse weather experiment: Use and evaluation of a citizen science tool for schools outreach. *Philosophical Transactions of the Royal Society A: Mathematical, Physical and Engineering Sciences, 374*(2077), 20150223.

61. Probert-Jones, J. R., (1962). The radar equation in meteorology. *Q. J. R. Meteorol. Soc.*, *88*, 485–495.
62. Rosenfeld, D., Wolff, D. B., & Amitai, E., (1994). The window probability matching method for rainfall measurements with radar. *J. Appl. Meteorol.*, *33*, 682–693.
63. Rosenfeld, D., Wolff, D. B., & Atlas, D., (1993). General probability-matched relation between radar reflectivity and rain rate. *J. Appl. Meteorol.*, *32*, 50–72.
64. Ryzhkov, A. V., & Zrnic, D. S., (1995). Comparison of dual-polarization radar estimators of rain. *J. Atmos. Oceanic Technol.*, *12*, 249–256.
65. Sachidananda, M., & Zrnic, D. S., (1986). Differential propagation phase shift and rainfall estimation. *Radio Sci.*, *21*, 235–247.
66. Schnädelbach, H., Rennick Egglestone, S., Reeves, S., Benford, S., Walker, B., & Wright, M., (2008). Performing thrill: designing telemetry systems and spectator interfaces for amusement rides. In: *Proceedings of the Sigchi Conference on Human Factors in Computing Systems* (Vol. 1, pp. 1167–1176).
67. Schonhuber, M., Lammer, G., & Randeu, W. L., (2007). One decade of imaging precipitation measurement by 2D-video-disdrometer. *Adv. Geosci.*, *10*, 85–90.
68. Scott, C. J., Bradford, J., Bell, S. A., Wilkinson, J., Barnard, L., Smith, D., & Tudor, S., (2016). Using the ionospheric response to the solar eclipse on 20 March 2015 to detect spatial structure in the solar corona. *Philosophical Transactions of the Royal Society A: Mathematical, Physical and Engineering Sciences, 374*(2077), 20150216.
69. Seed, A. W., Nicol, J., Austin, G. L., Stow, C. D., & Bradley, S. G., (1996). The impact of radar and rain gauge sampling errors when calibrating a weather radar. *Meteorol. Appl.*, *3*, 43–52.
70. Sekhon, R. S., & Srivastava, R. V., (1970). Snow size spectra and radar reflectivity. *J. Atmos. Sci.*, *27*, 229–267.
71. Seliga, T., & Bringi, V. N., (1976). Potential use of radar differential reflectivity measurements at orthogonal polarizations for measuring precipitation. *J. Appl. Meteorol.*, *15*, 69–75.
72. Sevruk, B., (1992). *Snow Cover Measurements and Areal Assessment of Precipitation and Soil Moisture* (Vol. 35, pp. 1–22). Operational Hydrology Report 35, WMO 749. World Meteorological Organization, Geneva.

73. Shaw, E. M., (1994). *Hydrology in Practice* (Vol. 1, pp. 1–20). Chapman and Hall, London.
74. Smith, P. L., (1984). Equivalent radar reflectivity factors for snow and ice particles. *J. Clim. Appl. Meteorol.*, *23*, 1258–1260.
75. Smoot, G. F., & Novak, C. E., (1969). Measurement of discharge by the moving boat method. *Water Resources Investigations* (Vol. 3, pp. 1–19). Chapter A.11. US Geological Survey, Washington DC.
76. Smyth, T. J., Blackman, T. M., & Illingworth, A. J., (1999). Observations of oblate 'hail' using dual polarization radar and implications for hail-detection schemes. *Q. J. R. Meteorol. Soc.*, *125*, 993–1016.
77. Stephenson, G., (1965). *Mathematical Methods for Science Students* (Vol. 1, pp. 1–19). Longmans, Green, Glasgow.
78. Straka, J. M., Zrnic, D. S., & Ryzhkov, A. V., (2000). Bulk hydrometeor classification and quantification using polarimetric radar data: Synthesis of relations. *J. Appl. Meteorol.*, *39*, 1341–1372.
79. Strangeways, I. C., (1999). Transmission of hydrometric data by satellite. Chapter 9. In *Hydrometry: Principles and Practices* (Vol. 2, pp. 245–264). Wiley, Chichester.
80. Strangeways, I. C., (2004). Improving precipitation measurement. *Int. J. Climatol.*, *24*, 1443–1460.
81. Testud, J., Bouar, E. L., Obligis, E., & Ali-Mehenni, M., (2000). The rain profiling algorithm applied to polarimetric weather radar. *J. Atmos. Oceanic Technol.*, *17*, 332–356.
82. Topp, G. C., Davis, J. L., & Annan, A. P., (1980). Electromagnetic determination of soil water content: Measurement in coaxial transmission lines. *Water Resour. Res.*, *16*, 574–582.
83. Ulaby, F. T., Moore, R. K., & Fung, A. K., (1981). *Microwave Remote Sensing* (Vol. 1–3, pp. 1–28). Addison Wesley, Reading, MA.
84. Ulbrich, C. W., (1983). Natural variations in the analytical form of the raindrop size distribution. *J. Clim. Appl. Meteorol.*, *22*, 1764–1775.
85. Villarini, G., Mandapaka, P. V., Krajewski, W. F., & Moore, R. J., (2008). Rainfall and sampling uncertainties: A rain gauge perspective. *J. Geophys. Res.*, *1*, 100–113. doi: 10.1029/2007JD009214.
86. Waldvogel, A., Federer, B., & Grimm, P., (1979). Criteria for the detection of hail cells. *J. Appl. Meteorol.*, *18*, 1521–1525.
87. WMO, (1966). *Measurement and Estimation of Evaporation and*

Evapotranspiration (pp. 95–115). Technical Note 83, WMO 201. World Meteorological Organization, Geneva.

88. WMO, (1976). *The CIMO Internation Evaporimeter Comparisons* (Vol. 449, pp. 1–20). WMO 449. World Meteorological Organization, Geneva.
89. WMO, (1982). *Methods of Correction for Systematic Error in Point Precipitation Measurement for Operations* (pp. 1–20). B. Sevruk. Operational Hydrology Report 21, WMO 589. World Meteorological Organization, Geneva.
90. WMO, (1985). *Casebook on Operational Assessment of Areal Evaporation* (pp. 1–18). Operational Hydrology Report 22, WMO 635. World Meteorological Organization, Geneva.
91. WMO, (1994). *Guide to Hydrological Practices* (Vol. 5, pp. 1–20). WMO 168, World Meteorological Organization, Geneva.
92. Young, T., (2018). 8.1-integrated network enhanced telemetry (iNET): Impact to the telemetry community for the ettc2018. *Proceeding-ettc2018*, pp. 155–161.
93. Zawadzki, I. I., (1984). *Factors Affecting the Precision of Radar Measurements of Rain* (Vol. 22, pp. 254–256). 22nd Conference on Radar Meteorology, Zurich, Switzerland, American Meteorological Society, Boston.
94. Zreda, M., Desilets, D., Ferre, T. P. A., & Scott, R. L., (2008). Measuring soil moisture content non-invasively at intermediate spatial scale using cosmic-ray neutrons. *Geophys. Res. Lett.*, *35*, L21402. doi: 10.10219/2008GL 035655.
95. Zreda, M., Shuttleworth, W. J., Zeng, X., Zweck, C., Desilets, D., Franz, T., & Rosolem, R., (2012). COSMOS: the COsmic-ray soil moisture observing system. *Hydrol. Earth Syst. Sci.*, *16*, 4079–4099.

Chapter 7

SatelliteBased Remote Sensing

CONTENTS

7.1. INTRODUCTION

Except for the reasonable endeavor which has been given to the advancement of the measurement systems, albeit by no means all, several of them are not viable yet for functional use comparable to groundbased systems. Undoubtedly, this situation will be changed dramatically by the current plans for the implementation of the new satellites. This chapter will outline the present, past, and future potential of the satellite systems. An increasing and significant subject area is the usage of satellite data for weather forecasting which will not be taken up here. For the development of the measurement systems having large coverage areas where traditional observations are absent or sparse, the opportunities have been provided by the advancement of satellite systems within low polar orbits, initiating by the television infrared observation satellites (TIROS) and after that Nimbus satellites, over the past 50 years. However, these methods endure restrictions set by weak temporal resolution because of the orbit geometry (having altitude of 600–800 km), measurements on particular areas are only being free after every 6 hours or 12 hours. This temporal resolution problem is rectified in the early 1970s, and late 1960s by the arrival of geostationary satellite orbits (having altitude of 36,000 km), METEOSAT, and GOES, but initially less spatial resolution is provided. In the last years, it becomes possible to have a high spatial resolution, less than 1 meter for infrared (IR) and visible measurements and for active microwave systems, less than one kilometer. Instruments are carried by early satellites which are only able to make calculations at IR, and visible wavelengths, though from the initial 1970s, radiometers are capable of serving data in many microwave bands, providing an ability to compute soil moisture, evaporation, rainfall, and transpiration, and to give atmospheric profiles of humidity and temperature. The range of frequencies is shown in Figure 7.1, which have been used.

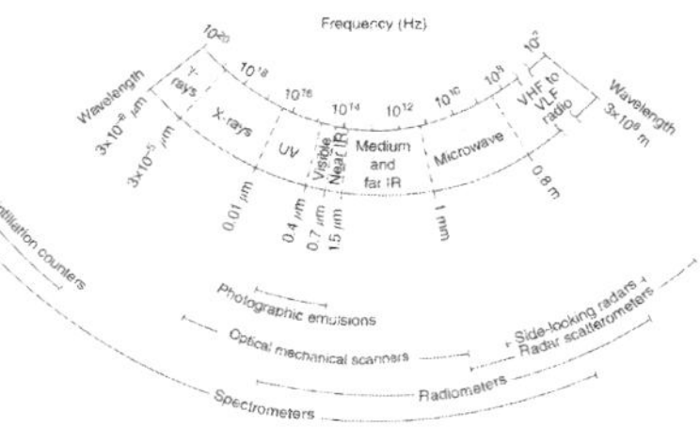

Figure 7.1. Satellite instrumentation connected with particular wavelength and electromagnetic spectrum.

Source: *https://onlinelibrary.wiley.com/doi/book/10.1002/9781118414965.*

For such satellite developments, most of the impetus has been the requirement to fulfil the needs of meteorology, albeit the Landsat, the ERS satellites and MODIS instrument possess empowered methods suitable to hydrology to be probed. The globally protracted repeatedly captured accumulation of the moderate resolution, spacebased, land remote sensing data is the Landsat program.

The website can be retrieved from earthengine.google.org. New means to watershed modelling is led by the presence of data having high temporal and spatial resolution (in contrast with some traditional measurements) (Van de Griend and Engman, 1985).

Now over broad areas, the parameters might be evaluated in detail, albeit the integral nature of spacebased and groundbased measurements has been acknowledged, especially on determining rainfall amount (Collier, 1985 and others).

Many various kinds of the instrument are employed by most of the recent satellite systems. However, generally, they might be classified into systems through IR, and visible wavelengths, and those employing microwave wavelengths, either actively (getting output of microwave by the satellite system that is retransmitted or reflected by the atmosphere or the surface, counting clouds), or passively (getting microwaves produced in the atmosphere, or over the Earth's surface).

For hydrological studies, currently, valuable missions are the MODIS and AMSRE aboard the AQUA platform, MODIS, and ASTER sensors on the TERRA platform, SMOS mission of ESA (where SMOS is abbreviated as Soil Moisture and Ocean Salinity), and also the upcoming Landsat series. TERRA is launched by NASA in 1999 which becomes functional in 2000.

For the understanding of the energy fluxes and the carbon, it is designed. MODIS for tracking large scale modifications in the biosphere, ASTER in consideration of land surface studies, and CERES in consideration of radiant and cloud energy monitoring are the instruments on this satellite.

The AQUA mission is launched by NASA in 2002 which targets to collect data over the Earth's water cycle. The 6 instruments are carried by this mission within an orbit of 705 km, and counts AMSRE, that is not an active microwave radiometer for frequencies in the range of 6.90 to 80 GHz. From 5.40 to 56 km is the range of the spatial resolution. A scientist, Jong et al. (2008), discussed more details of the AMSRE and ASTER instruments.

7.2. ELECTROMAGNETIC RADIATION'S SURFACE SCATTERING

All objects emit thermal radiation at temperatures up to definite zero. This radiation is exposed by many non-active remote sensing systems. Normally, its emission will be distributed by a hot object at a range of directions, and wavelengths. Real materials do not act being a black body, hence, *emissivity* ε is presented which denotes the real radiance of a body by temperature T to the value of the black body. Usually, emissivity is based upon wavelength. Thus, the spectral radiance of the real object is

$$L_\lambda = \varepsilon(\lambda) L_{\lambda p} \qquad (1)$$

where; subscript p indicates Planck; and $L_{\lambda p}$ signifies the black body radiance. A good emitter body (e.g., increased ε) should also be a strong radiation absorber. *Reflectivity* is $r = 1 - \varepsilon$, thus, emissivity is situated within 0 and 1.

The surface's thermal emissivity is associated with its reflectance characteristics. A well-imitating radiation beam is shown in Figure 7.2, where flux density F(W m^{-2}) is evaluated at right angles to the course of propagation occurrence at angle θ_0 on a surface. An element of such radiation is dispersed in a direction given by the angle θ_1, into the steady angle $d\Omega_1$. Usually, θ_0 is known as the incidence angle, while its complement $(\pi/2)-\theta_0$ is called the depression angle.

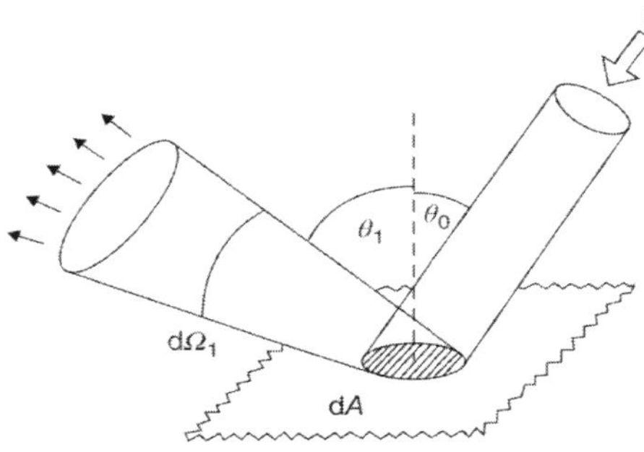

Figure 7.2. At angle θ_0, starting from flux density F, radiation is incident over an area dA and then spread in a direction of θ_1 into solid angle $d\Omega_1$. For clarity, the azimuthal angles φ1 and φ0 are excluded.

Source: *https://onlinelibrary.wiley.com/doi/10.1002/9781118414965.ch14.*

For the radiance of out direction of the surface E consequent to its illumination L_1 (W $m^{-2}sr^{-1}$) for the direction (θ_1, φ_1), then the *bidirectional reflectance distribution function* (BRDF) R is expressed as:

$$R = L_1 / E \quad (2)$$

The subscript 1 shows the direction of scattering. The reflectivity r is indicated as the *albedo* and is linked with the emissivity as given below:

$$r = 1 - \varepsilon \quad (3)$$

Conventional values of albedo are given in Table 7.1.

Table 7.1. Conventional Values of Albedo

Material	**Albedo (%)**
Cloud (high)	30–85
Cloud (low)	5–65
Crops (green)	5–15
Forest	5–10
Global average	≈35
Granite	30–35
Grass	5–30
Ice	25–40
Lava	15–20
Limestone	35–40
Sand	20–40
Snow (fresh)	75–90
Snow (old)	45–70
Soil	5–30
Urban areas	5–20
Water (spontaneously occurring)	1–10
Water (uncontaminated or pure)	2.0

There are two fundamental methods to compute BRDF (Rees, 1993):

1. **Facet Model:** Where the surface is imaged to be made up of planar facets all are tangential to the real surface. Moreover, the facets should be greater than single wavelength side to side.
2. **Small Perturbation Method:** Where the surface interaction with the incident field is employed to evaluate the outgoing field for the surface vicinity.

The reflectance is considered from the rough surfaces as based on examining directions and the incidence but not the wavelength. Nevertheless, it is illustrated in Figure 7.3 which is not suitable. Figure 7.3 demonstrates the simplified spectral albedos by different materials in the near, and visible IR wavebands. For instance, it is obvious that working at around 1.20 μm (i.e., near-IR) only leads to struggle in segregating within a total material. If 3 or 4 channels are utilized within 0.5 and 1.20 μm, then this issue is minimized.

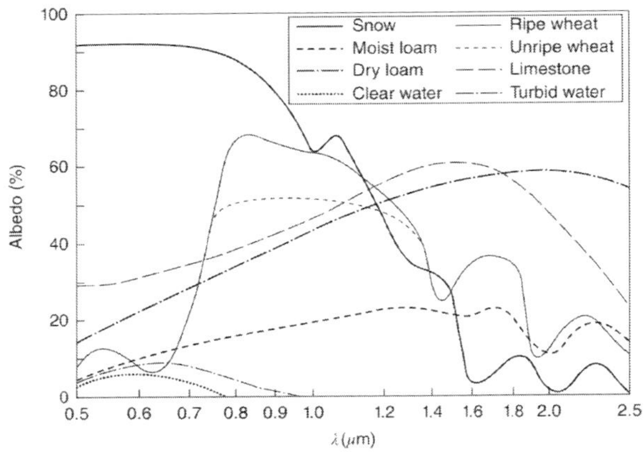

Figure 7.3. Conventional spectral albedos of different materials in the near, and visible infrared.

Source: *https://onlinelibrary.wiley.com/doi/10.1002/9781118414965.ch14.*

A sudden change is appeared in the vegetation reflectance at around 0.70 μm, that is employed by various methods for getting the total vegetation as shown in a figure (Harris, 1987). In the near and visible IR, the wavelength-independent dispersing events situated at the interfaces between air and ice causes the uniform and high reflectance of snow. Thus, snow can be identified, under favorable conditions, when lighted just from moonlight (Foster, 1983). As compared to the reflectance properties of vegetation, that of the rock are little distinctive, and the reflectance characteristics of soil are based on the kind of rock from where the content of water and vegetable matter, and the soil is deduced.

Figure 7.4 shows the emissivity of different materials within the thermal IR region. Figure 7.4 clarifies that for the identification of different materials, multifrequency measurements in passive microwave radiometry in the range of 1 to 36 GHz are very valuable.

7.3. SYNERGY OF ELECTROMAGNETIC RADIATION ALONG WITH THE ATMOSPHERE

Two journeys for systems, or one journey by the atmosphere are made by the radiation that identify reflected radiation. Both scattering and absorption are experienced every time when radiation moves by the atmosphere. *Attenuation* is referred to as the total of these two energy losses. The atmosphere's refractive index also differs from unity. All of these effects are based upon wavelength.

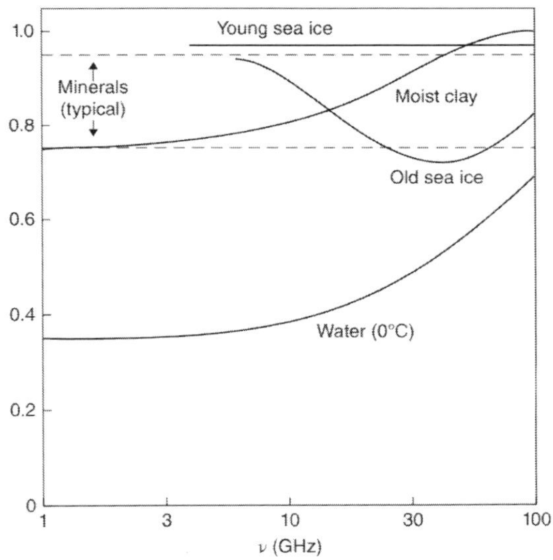

Figure 7.4. At a range of frequencies ν, emissivity of microwave ε in regular direction of different materials.

Source: *https://onlinelibrary.wiley.com/doi/book/10.1002/9781118414965.*

The range of attenuation is expressed logarithmically in decibels. The range to which the intensity of radiation invades through it is provided by the *optical depth* τ (sometimes known as the *opacity*), and is attenuated by a given layer of material, and is expressed as:

$$\tau = \ln(I_1 / I_2) \quad (4)$$

where; I_1 denotes the incident intensity (i.e., unattenuated); and I_2 indicates the output intensity (i.e., attenuated). At 0.55 μm wavelength, the sum of the optical depth of the atmosphere, under normal conditions, is around 0.3. Hither, 0.2 is participated by particulate matter, and 0.1 by atmospheric gases. For a vertical ray, total atmospheric attenuation is shown in Figure 7.5. A study, by Rees (1993), contains a detailed description of the structure of Figure 7.5. It is considered to be the ground for the remote sensing methods employed to sound the atmosphere. Parameters like pressure and temperature can be evaluated as a function of height, and the height of a particular layer may be differed by controlling the observation frequency (refers to, for instance, Harries, 1994).

7.4. INFRARED (IR) AND VISIBLE DATA

Estimation of a variety of parameters is allowed to be made by satellite data at visible frequencies, that are significant to hydrological modelling, e.g., the part of polluted area or of water surfaces, and the part of impervious area within a river catchment (McGinnis et al., 1979). There might be the analysis of the flood inundation areas as well, where given data resolution is enough (Ramamoorthi and Rao, 1985). Actually, in some areas, it is also possible to conclude presumably water resources that are underground by such data (Salman, 1983). However, IR, and visible data are most generally used to constitute approximates of soil moisture, snowfall, and rainfall. These approximate are created by corresponding the microwave radiation diffused by a surface, thermal IR radiation coming out of a surface, or the Earth's surface reflectivity in the visible wavelengths or clouds, to independent calculations of the parameter.

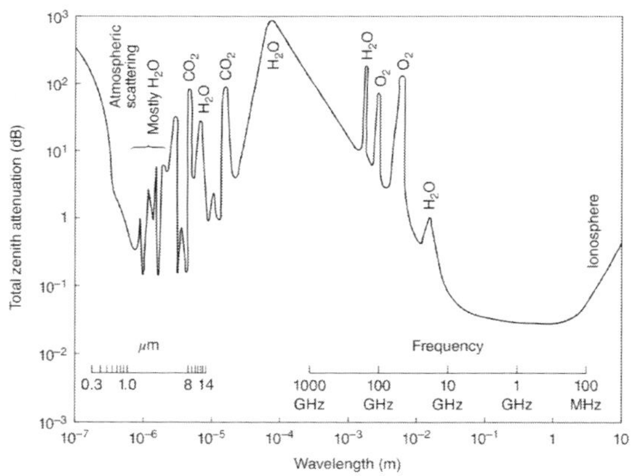

Figure 7.5. Complete zenith attenuation on normal conditions over the atmosphere (or simplified).

Source: *https://link.springer.com/chapter/10.1007/978-3-319-44431-4_13.*

The relationship within the absolute temperature *T* of the radiating surface, and the wavelength of maximum emitted radiance λ_{max} is expressed by *Wien's displacement law*, describes that for a *black body,* the product of wavelength and absolute temperature is constant (ideal emitter at each wavelength):

$$\lambda_{max} T = w \quad (5)$$

where; *w* signifies the Wien's displacement constant.

The *Stefan-Boltzmann law* provides the *infrared radiant emission* which relates the 4th power of the absolute temperature to the total emittance of radiant from a black body:

$$W = \varepsilon \sigma T^4 \quad (6)$$

where; σ represents the Stefan-Boltzmann constant; *W* signifies radiant power for each unit area; *T* represents the absolute temperature (K); and ε denotes the emissivity of the surface. Of every typical natural surface, water is presumably a black body, and there may be a variance of emissivity from 0 to below 1 and based on wavelength. The material's resistance to change in temperature refers to the *thermal inertia P*, given by:

$$P = (kx\rho)^{1/2} \quad (7)$$

where; ρ indicates the density; *k* represents the thermal conductivity; and *c* indicates specific heat.

The atmosphere changes the thermal IR temperature which is calculated by a satellite. The radiation is evaluated while considering down vertically is partially composed of reflected radiation coming out of the sky in the clear sky case. The abrupt changes can occur in the radiant emission from a surface when clouds exist.

7.4.1. Precipitation

There are two kinds of methods that are presented for the estimation of rainfall, recognized as *life history* and *cloud indexing* methods.

The first method that has been presented for the estimation of rainfall is cloud indexing. A cloud index or rainfall coefficient is measured by the attributes of the satellite cloud field explained by IR or visible images, for instance, texture or brightness. After that, these indices become analogous, through regression equations, to rain gauge measurements of rainfall (Follansbee, 1973; Barrett, 1970, 1976; Follansbee and Oliver, 1975).

Such methods are most favorable for rainfall for periods of months or days, partly as by polar orbiter platforms, satellite data are only available after every 6 hours or 12 hours. Latently, for use along with geostationary data, alike methods have been introduced (Delbeato and Barrell, 1985), although generally, they only operate good for convective cloud, because of the against to frontal cloud. The derived relationship is shown in Figure 7.6, over southeast Australia within cloud top temperature and 90 minutes rainfall totals. Pattern recognition techniques through radiance or textural features are being determined as well (refers to for instance Wu et al., 1985).

Scientists are capable to determine the rainfall formed using convective clouds by the total area of the clouds, and the amount of change of that certain area Stout et al. (1979). Which refers to the volumetric rain rate towards a specific cloud is expressed as:

$$R_v = a_0 A_c + a_1 \frac{dA_c}{dt} \tag{8}$$

where; dA_c/dt represents the rate of change of cloud area; A_c specifies the area of the cloud; and a_1 and a_0 denotes empirical coefficients. Satellite images over frequent time intervals are required by this type of approach called life history technique, which can be given by geostationary satellites only.

Another scientist has produced the variant of this technique. <Stout et al. (1979) is the technique, and presented by Griffith et al. (1976, 1978)> (Scofield and Oliver, 1977a, b; Scofield, 1984; Woodley et al., 1980). In the variant method, high cloud brightness (i.e., low cloud top temperature) favors the precipitation, and cold cloud top temperatures, and merging and growth of the clouds favor heavy rainfall. A scientist has proposed advancements to cope more sufficiently with the storms with warm tops Scofield (1981). Another researcher has been studied for the addition of textural and spectral data from geostationary satellite data towards the evolutionary data and get to present valuable improvements in the tropics Martin and Howland (1986). Figure 7.7 shows the type of techniques that indicates the classical instance of a rainfall field deduced from polar and geostationary orbiting satellites both by IR satellite data.

Some scientist found that the least observed black body temperature through the life period of a storm is associated with the utmost volumetric storm rainfall, and top ascent rates of thunderstorm are associated with the utmost reflectivity of storm radar Negri and Adler (1981).

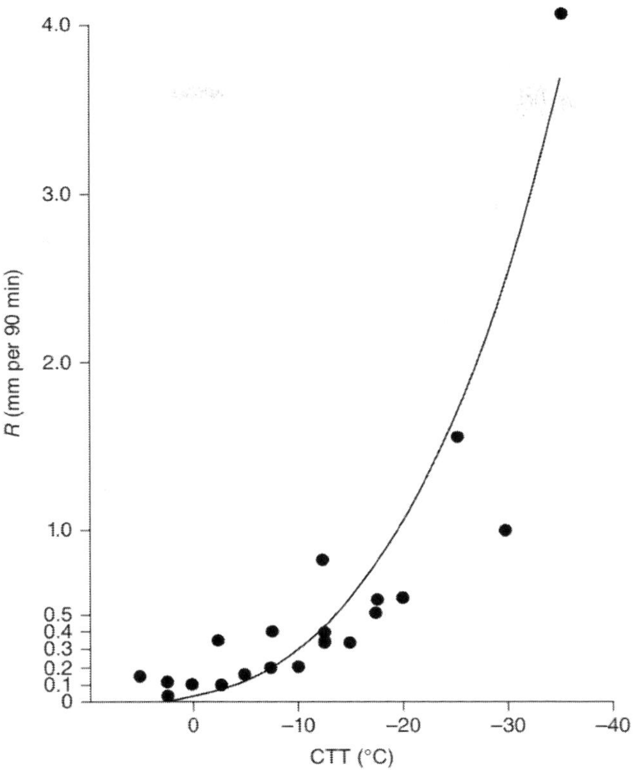

Figure 7.6. Noticed cloud top temperatures CTT on southeast Australia and surface rainfall totals 90-minute R for the cloud in the convective region of extratropical cyclones or cumuliform cloud; the subjects are obtained from many studies.

Source: *https://www.researchgate.net/publication/316325743_Hydrometeorology_in_the_Urban_Environment.*

Generally, this method only gives adequate estimates of convective rainfall. Nevertheless, it is summarized by a researcher that for regular rainfall estimation, the Griffith-Woodley method is unnecessarily complexed Negri et al. (1984). A simple process has been proposed where the area-time integral (ATI) about cloud areas at the period of storm is connected with the sum of rain volume Doneaud et al. (1985). In literature, there is a discussion of the studies that are using this technique by radar data. Again, this method appears to operate well just in consideration of convective rainfall.

7.4.2. Snow Depth

Seasonal transformations in snow cover can be tracked routinely from thermal IR and satellite visible pictures (Rango and Martinec, 1979; Rango et al., 1977; Dey et al., 1979, 1983), even though the usage of multispectral data is required by the accurate categorization of the types of snow and distribution. Scientists have reviewed the usage of such techniques in the USSR (Rango and Itten, 1976; Rango and Peterson, 1980; Deleur, 1980).

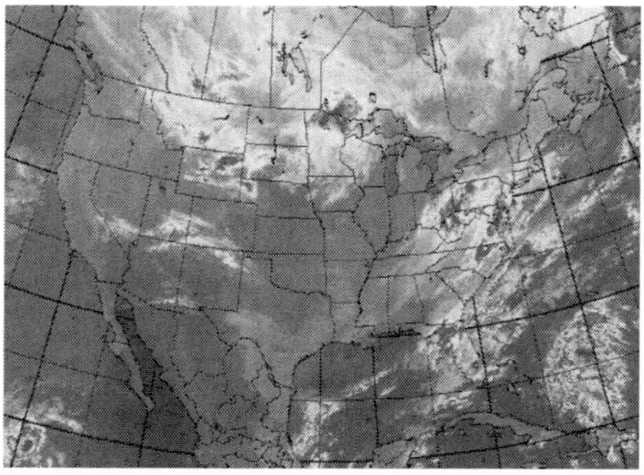

Figure 7.7. Rainfall predicts by infrared satellite data, 1630 GMT 8 October 2012: The photographs presented of US satellite indicates infrared (IR) figures.

Note: Coldest (i.e., highest) clouds are shown in shades of purple, red, and yellow; warmest (i.e., lowest) clouds are displayed in white. Imagery is acquired from the couple of US polar orbiter (POES) satellites, and the METEOSAT and GOES geostationary satellites. The revolution of the POES satellites around the Earth is 14 times every day at a height of around 520.0 miles (870.0 km)

Source: *https://link.springer.com/chapter/10.1007/978-3-319-67047-8_2.*

Measurement methods dependent on the usage of IR or/and visible pictures are restricted to discerning the surface circumstances of the snow (Martinec and Rango, 1981). Hydrologists must evaluate the potential runoff because the snow melts. The usage of the data of high-resolution satellite along with *in situ* surface measurements in the hydrological analysis of snowmelt in consideration of hydroelectric production is exhibited by Odegaard et al. (1980). Masking formed by clouds is one of the bottlenecks while evaluating the range of snow cover by a satellite. Nevertheless, the distinguishing of snow cover and clouds are possible through a range of spectral bands.

7.4.3. Evapotranspiration (ET) and Soil Moisture

The connection of the reflectance of soil spectral to the water content is based on many variables' inclusion with the water content, e.g., the organic content and soil texture. Thus, the usage of reflected energy is impractical, either IR or visible.

However, the evaluation of the difference within the temperature within a crop and air temperature or the diurnal extent of surface temperature might be connected to soil moisture (Wetzel et al., 1984; Idso et al., 1975a, b). For the measurement of the diurnal temperature extent, an empirical method has been proposed by the evaluations from the Heat Capacity Mapping Mission between 1978 and 1980 (Carlson et al., 1981). Some scientists have determined a similar method for the measurement of evapotranspiration (ET) deduced through surface temperatures (Klaassen and Van Den Berg, 1985; Gurney and Camillo, 1984).

7.5. MULTISPECTRAL DATA

In the previous section, the usage of sequences of images, or the single images, of either IR or visible data has been already discussed. But with the possible anomaly of the measurement methods of rainfall, for functional implementation, techniques have not been verified sufficiently reliable. Improvements can be found by the combined consumption of data from many spectral bands.

7.5.1. Precipitation

Data about temperature is provided by IR sensors, and hence not directly over the altitudes of the tops of clouds. Whereas, information about the depth of clouds and their composition and geometry are provided by the visible sensors. The combination of such data allows identification of high cloud tops connected with deep clouds which somehow results in symbolic rainfall (Barrett and Martin, 1981). Some of the scientists have done the initial work on this kind of technique (Dittberner and Vonder Haar, 1973; Lethbridge, 1967; Reynolds et al., 1978). But challenges facing by instrument calibration, the time difference between illumination geometry and images, and registration errors between IR and visible images are the reasons that the early work is proved to be less promising than the scientists had expected. Scientists have also presented ways to overcome such challenges (Lovejoy and Austin, 1979a).

Many studies utilizing models of convective procedures targeted to cope with the challenges from a physical, instead of a statistical view point (Gruber, 1973; Wylie, 1979).

Rain areas are defined by these methods. But more challenging is the evaluation of rainfall amounts, and groundtruth data that is found from radar might be required for the analysis of the satellite information on the reliable valuation of rain area, or valuation making of rain amounts. A scheme is presented that employs radar data to determine the rain areas and no rain for calibration and after that utilizes the satellite IR and visible data just to delineate rain areas; there is no production of the quantitative estimates (Tsonis and Isaac, 1985).

7.5.2. Cloud Recognition

Data with the cloud top temperature, and high spatial resolution is provided by the recent generation of satellites. Unluckily, such data is not enough by itself to enable the automated investigation of various cloud types. The problems can be alleviated by the combination of information from various spectral bands.

A technique is evolved employing information from 3 AVHRR spectral channels is given below (Liljas, 1982):

Visible (VIS)	0.57–0.68 μm	(NOAA6, 7)
–	0.54–0.90 μm	(TIROSN)
Infrared (IR)	10.50–11.50 μm	–
Near-infrared (NIR)	0.725–1.2 μm	–

The method is same as to that one presented by another scientist (Barrett and Grant, 1978), but is dependent upon the supposition that various cloud types, water surfaces and land surfaces possess a variety of radiational characteristics in several areas of the electromagnetic spectrum. Figure 7.8(a) illustrates the derivation of signatures from various cloud types inside the boxes within threedimensional intensity space. The method has experimented over southern Scandinavia with promising outcomes.

7.5.3. Snow

In previous sections, the challenge of segregating snow and cloud cover on IR and visible images are already discussed. Following is the data that is given by the thematic mapper (TM) over Landsat 4 and 5 in seven wavelengths:

- Band 1: 0.44–0.53 μm;
- Band 2: 0.51–0.61 μm;
- Band 3: 0.62–0.68 μm;
- Band 4: 0.75–0.91 μm;
- Band 5: 1.55–1.76 μm;
- Band 6: 10.41–12.51 μm;
- Band 7: 2.07–2.30 μm.

Bands 1 to 5 and band 7 have 30 m spatial resolution and are in the near-infrared, visible, and middle IR wavelength zones. The resolution of band 6 is 120 m and is a thermal IR band. The conventional spectral reflectance curve is shown for snow in Figure 7.9, which demonstrates the saturation standards for the near-infrared and visible TM bands. Band 5 may be utilized to discriminate within snow and clouds because the snow reflectance declines remarkably at around 1.50 μm wavelength, on the other hand, there is a high reflectance of clouds.

7.6. PASSIVE MICROWAVE APPROACHES

The microwave radiation (about 10 to 0.10 cm wavelength) emitted thermally from an object is known as its brightness temperature T_B. The microwave radiation that is radiated from black body and its physical temperature T is proportional, however, as concerning IR emission, most real objects or surfaces radiate just a part of the radiation, that a black body would radiate at equal physical temperature.

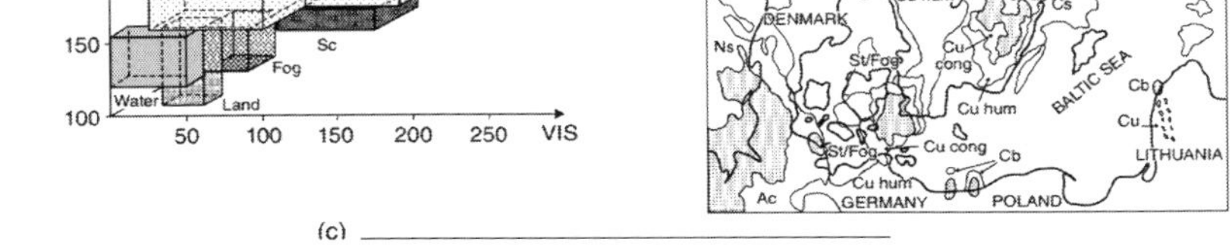

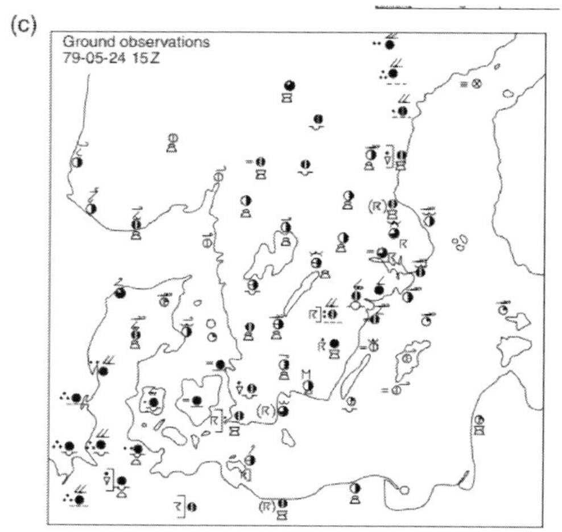

Figure 7.8. Demonstrating (a) a categorization in a 3dimensional intensity space of fog, cloud, water, and land; (b) cloud examined from this category on 1500 UTC 24 May 1975 on southern Scandinavia; and (c) the weather noticed around 1.5 hours subsequent to the satellite passage.

Source: *https://www.amazon.com/Hydrometeorology-Advancing-Weather-Climate-Science/dp/1118414977.*

46° Solar zenith angle MSS 10° Solar zenith angle MSS	46° Solar zenith angle TM 10° Solar zenith angle TM

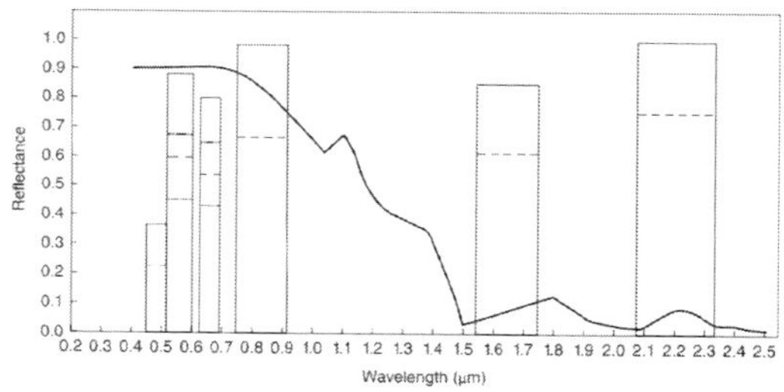

Figure 7.9. For snow (O'Brien and Munis, 1975), classical reflectance curve displaying saturation standards for near and visible infrared bands of the multispectral scanner (MSS), and thematic mapper (TM).

Source: *https://search.library.uq.edu.au/primo-explore/fulldisplay?vid=61UQ&search_scope=61UQ_All&tab=61uq_all&docid=dedupmrg1169320047&lang=en_US&context=L.*

This part represents the emissivity of the object ε, provided, for wavelengths of microwave, by:

$$T_B = \varepsilon T \quad (9)$$

With the measurement of the physical temperature by some independent ways, and with the measurement of the brightness temperature with the help of a radiometer attached to a satellite, the emissivity might be obtained. But, with the variation of wavelength, the emissivity generally differs, and hence by determining the brightness temperature at higher than one microwave wavelength, either the estimation of more than one physical property or finding the physical temperature by the radiometric measurements is possible. Approaches for inferring such characteristics from a set of brightness temperature measurements over multiple path angles or wavelengths are called *inversion techniques*.

For accessing surface and atmospheric parameters, inversion techniques have been used, parameters like rainfall and humidity and temperature profiles. A thorough review of such procedures is presented by which the given below is adapted (Ulaby et al., 1986). Suppose that the brightness temperature $T_B(\lambda_i)$ is linked linearly to the needed parameter denoted by $g(z)$ function, where z represents the vertical distance starting from the radiometer, therefore,

$$T_B(\lambda_i) = T_{BGD}(\lambda_i) + \int w(\lambda_i, z) g(z) dz \quad i = 1, 2, ..., n \quad (10)$$

where; $w(\lambda_i, z)$ represents the weighting function of $g(z)$; T_{BGD} denotes the temperature of background radiometric related to the surface of the Earth. When observations are formed at an angle except for the nadir or zenith (vertically below the satellite) based upon the platform, thus, z is interchanged to $z \sec \theta$. For a non-linear relationship, it is quite probable to consume a similar model to that expressed by Eqn. (6), albeit the weighting function becomes more complicated. Usually, the nonlinear relationships are for wavelengths nearer to the IR zone, and for microwave wavelengths, there is a linear relationship. Usually, T_{BGD} is certainly variable as the emission from the surface of the Earth differs temporally and spatially. Certainly, in evaluating snow depth or soil moisture content, for instance, the consumption of the isothermal Eqn. (5) is not practical. Account has to be taken of a class's radiative function expressed in Eqn. (6), that explains the transmission of radiation at a given point inside the material constructing the surface layer of the Earth towards the interface of the ground-atmosphere.

For the emissivity calculation, a weighted mean of the physical temperature at the extent of the surface layer *T* is utilized:

$$T_B = \varepsilon T \quad (9)$$

The depth recognized being the *skin depth* is based on the surface layer's material, from where there is an emission of the 63.0% of the radiation. This depth is numerous meters for dry snow, numerous centimeters for sea ice that is many years old, many millimeters for soils, water, and firstyear sea ice at 1.56 cm wavelength (Zwally and Gloersen, 1977).

7.6.1. Precipitation

For the measurements of precipitation, the usage of the electromagnetic spectrum's microwave zone has the upper hand on IR and visible methods of identifying the precipitation by its own, amidst clouds being transparent partially or fully.

For the measurement of the natural radiation within the microwave region, one of the earliest instruments to be utilized meteorologically is ESMR5 (electrically scanning microwave radiometer) that has been presented in 1972 on NIMBUS5. The simple contrast of the NIMBUS7 SMMR (scanning multichannel microwave radiometer) metaphors working at 19.34 GHz (i.e., 1.56 cm wavelength) along with metaphors from IR, and visible radiometers of the wavelength, traditional meteorological observations and groundbased radars are set up over water a corporation of regions of comparably warm brightness temperature along with the zones of rainfall (Wilheit et al., 1973, 1977; Theon, 1973). Wilheit et al. (1977) published curves of rainfall rate and 19.35 GHz brightness temperatures with the help of the model depends on the idea of several optically thin layers fixed to the top with a fluctuating freezing level. Many problems have been identified in this work, particularly the nonlinear relationship between brightness temperature and rainfall rate, and in addition to that the rarely uniform rainfall over the complete instrument's field view (Spencer et al., 1983). A researcher has presented the advancements to the method, inclusive of the usage of both horizontal and vertical polarization concerning over land measurements Wilheit (1975). However, because of the variability of the emissivity, it is much more troublesome to measure over the land, and thus the background radiation is evaluated through instrument of satellite. An approach has been proposed which depends upon the study of radiometric information acquired through the NIMBUS7 SMMR over a frequency of 37 GHz (i.e., 0.81 cm wavelength) in an effort to overwhelm the challenges of nonlinearity that are discussed above, and confusions presented within emissions because of the rain and those because of nonprecipitating cloud (Spencer, 1986). Dual polarized data is provided by this radiometer system at five various wavelengths (i.e., 1.43, 0.81, 2.00, 1.66, and 4.53 cm), and this system can be utilized to identify a broad range of ice and snow features including rainfall. It is illustrated in Figure 7.10 that the difference within how the brightness temperatures of the vertical and horizontal polarized radiation on 37.0 GHz is connected to the average brightness temperature and rate of rainfall concerning convective rainfall at the United States. It is suggested by a scientist that 37.0 GHz can offer the most compromise for evaluating rainfall at oceans within the ability of cloud penetration and a conveniently identifiable precipitation signal (e.g., at 19.36 GHz) (Spencer, 1986). Nevertheless, it is observed that as compared to the strongness of radiated signal concerning light precipitation over 19.0 GHz, the dispersing signal over 37.0 GHz is not as powerful. Thus, 37.0 GHz strategy might be exceptional for noticing heavy convective precipitation, and observations over 19.0 GHz may present better to evaluations of light precipitation. Plans have been designed to add on a screening microwave radiometer over the US Polar Platform Space Station (McElroy and Schneider, 1985), a scheme is named the Columbus Project in Europe.

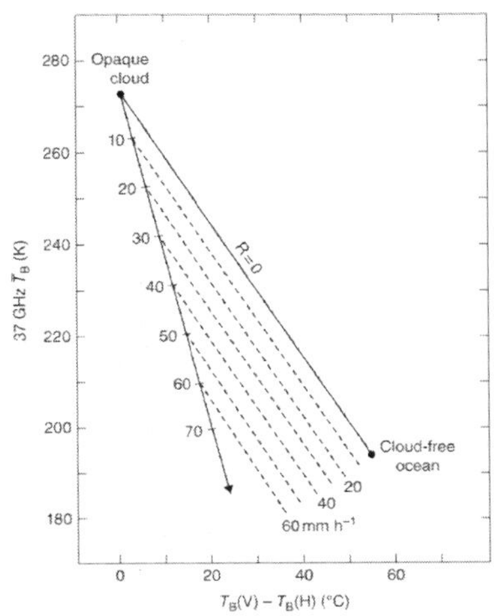

Figure 7.10. Bipolarization (vertical/horizontal) graph of rainfall rates along with the noticed relationship within SMMR brightness temperature TB at the time of the summer and radar rainfall rates in the United States.

Source: *https://www.kobo.com/gr/en/ebook/hydrometeorology-3.*

But, in 1991, Columbus has been abolished. In 1984, the US space station, named FREEDOM, has been declared, but is not manufactured; it has progressed as the International Space Station, whose first unit has been released in 1998. For the past 10 years, meteorological attempts have been performed, albeit routine meteorological calculations have been taken from another spacecraft. The microwave brightness temperature information of 25 years that is given by the SSMIS and SSM/I tools aerial on the DMSP satellites is described (Berg et al., 2012).

7.6.2. Global Precipitation Climatology Project (GPCP)

Applying input data methods and sets from a variety of investigators, the GPCP is an international scheme. The availability from January 1979 to the present, of a universal comprehensive monthly analysis of surface precipitation at a 2.50° longitude × 2.50° latitude resolution is accessible. Precipitation estimations from low-orbit satellite microwave data (accessible since 1987), surface rain gauge statistics, and geosynchronous IR figures are combined in this fused analysis. Version 2 of the data in which new figures have been defined as described by researchers (2003), while version 1 was detailed by Huffman et al.

7.6.3. Global Precipitation Measurement Mission (GPM)

The nextgeneration worldwide inspections of rain and snow are given by an international system of satellites known by the name; The global precipitation measurement mission (GPM). Its idea focuses on the measurement of precipitation through a highly developed radar/ radiometer structure by utilizing a "core" that transports this radar/radiometer, picking up on the achievement of the TRMM, otherwise known as "tropical rainfall measuring mission (TRMM)." The unification of precipitation measurements from a collection of operational satellites and research will also be done by acting as a reference standard (Figure 7.11). Many space agencies exist within GPM, which is an origination by the JAXA; Japan Aerospace Exploration Agency and NASA, being a global descendant to TRMM, a few including it are; US National Oceanic and Atmospheric Administration (NOAA), Center National d'Études Spatiales (CNES), European Organization for the Exploitation of Meteorological Satellites (EUMETSAT), Indian Space Research Organization (ISRO), and others. Tanegashima Space Center, Japan was the place where The GPM Core Observatory was commenced at 1337 EST on 27 February 2014.

7.6.4. Snow Depth

The microwave discharge from a substance is governed by the electrical characteristics which create a substance's specific dielectric constant (which is a measurement of the response to an operating electrostatic field).

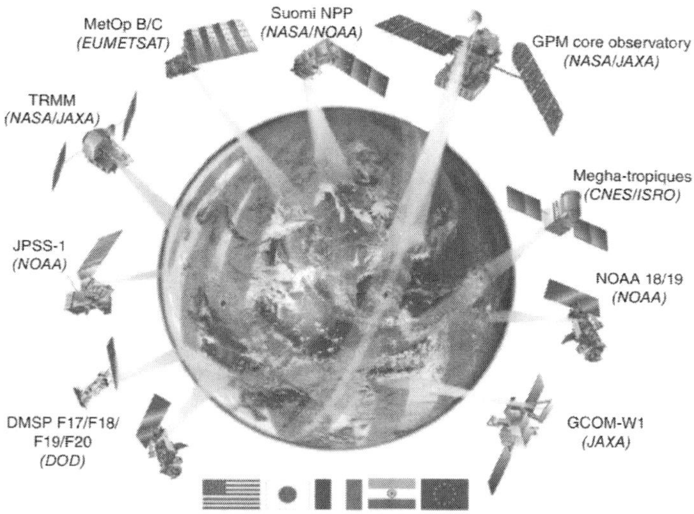

Figure 7.11. The constellation of global precipitation mission (GPM).

Note: French Water Cycle, MeghaTropiques, and Energy Budget in the Tropics (2011-); Global Change Observation Mission, GCOMW1, Suomi NPP (2012-); (NPP) National Polar-orbiting Partnership (2011-); NOAA Joint Polar Satellite System, JPSS1 (2017-).

Source: *https://gpm.nasa.gov/education/images/gpm-constellation-satellites.*

The echo of an incident microwave transmission or dispersal of the native microwave discharge can be caused by variations in the dielectric constant within a substance, which is a role of frequency and temperature. Thus, within the snowpack, the existence of water, ice, and air must be heeded while considering the electrical features of snow. Models derived on haphazardly spaced scattering spheres have been established due to it as can be seen by Chang et al., 1976. It has been resolved that enhanced snow cover data could be offered by microwave sensors over optical sensors by examining the response of the above-mentioned kind of model to microwave employing SMMR figures. Numerous studies noted by other researchers defend this work. The snow depth versus brightness temperature at two diverse wavelengths and two diverse polarizations are displayed in comparison by a researcher in Figure 7.12 (Rango et al., 1979). In comparison to snow depth at 1.55 cm, at a wavelength of 0.81 cm, the temperature brightness demonstrates a greater sensitivity.

It was concluded that it is sensible to obtain relationships particular to a region of interest as the relationship amongst brightness of temperature and snow depth differed from one region to another. However, more extensively applicable relationships may be provided by the utilization of dual-polarization at two dissimilar frequencies. Formulae were developed for water equivalent, snowmelt, snow cover, and snow depth by Kunzi et al. (1982).

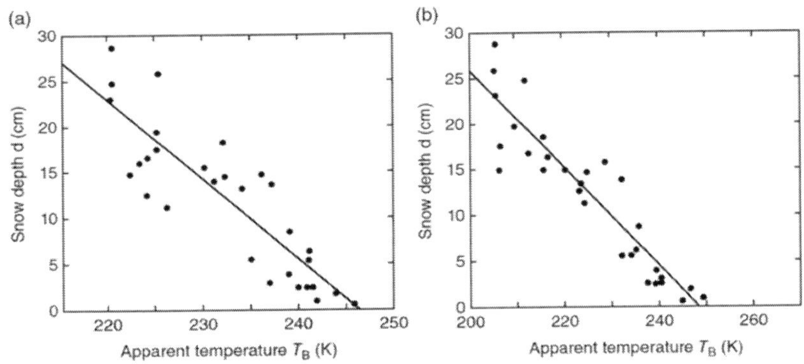

Figure 7.12. Snow depth versus brightness temperature calculated by (a) the NIMBUS-5 ESMR at horizontal polarization, 1.55 cm; and (b) the NIMBUS-6 ESMR at vertical polarization, 0.81 cm.

Source: *https://ntrs.nasa.gov/api/citations/19800019278/downloads/19800019278.pdf.*

7.6.5. Temperature of Sea Surface and Sea Ice

NIMBUS5's (Salomonson and Gloersen, 1975) ESMR data was employed to make the initial satellite examinations of the sea ice and temperature of sea surface. Remarkable microwave pictures of the Arctic and Antarctica in both winter and summer were presented by Zwally and Gloersen (1977). Figure 7.13 displays current examples of sea ice climatology.

When viewed from a satellite, the evident sea surface temperature T_S at angle θ is presented by:

$$T_S = T_B e^{-r\sec\theta} + \Gamma T_D e^{-r\sec\theta} + T_U \quad (12)$$

where; Γ is the reflectivity of the surface; T_D and T_U are the upward and downward atmospheric brightness temperatures, τ stands for the total atmospheric opacity and T_B stands for the regular sea surface's brightness temperature. Input from 1st-year ice, seawater, and multi-year ice (of a dissimilar arrangement, because of refreezing after melting, to 1styear ice) are part of $T_{B.}$ The degree to which the field of view of the satellite consists of open water controls the exactness of estimates of ice concentration. It also depends upon the degree to which it is likely to adjust the measurements utilizing thermal IR estimations in cloud-free areas of the ice temperature. Sea surface temperature information has been collected by the NOAA polar-orbiting satellites (POES) for more than 22 years, and this data is currently an essential portion of systems of ocean forecasting with sea ice data (Martin et al., 2007).

7.6.6. Evapotranspiration (ET) and Moisture in Soil

The minimization of atmospheric outcomes for measuring soil moisture is done by using longer wavelengths, and the benefit of being able to infiltrate vegetation rests with microwaves.

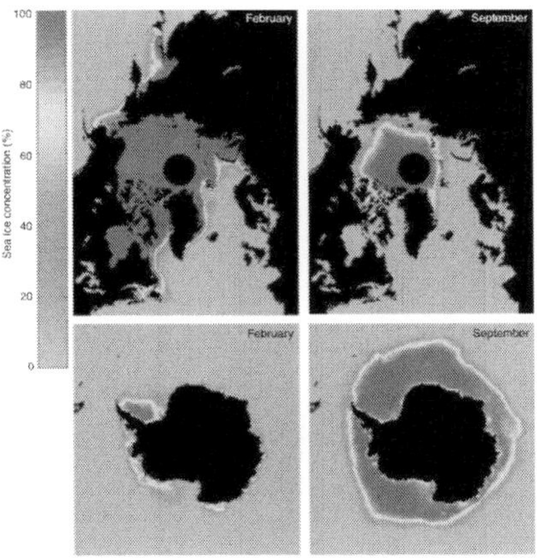

Figure 7.13. Antarctic (bottom panels) and Arctic (top panels) sea ice concentration climatologies from 1970's to 2002: Approximated seasonal winter minimum and maximum (February in the Antarctic and September in the Arctic) stages.

Source: *http://www.antarcticglaciers.org/glaciers-and-climate/changing-antarctica/antarctic-sea-ice/.*

How the generation of thermal microwave discharge arising out of soils takes place inside the soil volume is defined by Jackson et al. (1981), also the temperature of the soil and relying on the dielectric features (chiefly moisture in the soil's case), the quantity of energy at a point. Along the track of propagation, the dielectric gradients (soil moisture) influence the energy as it spreads upwards through the soil. Reduction of the energy takes place at the surface boundary due to the effective transmission coefficient (the emissivity); governed by the dielectric properties of the soil close to the surface.

The skin depth stated earlier is the soil level that governs this transmission coefficient. For a wavelength of 21 cm, it is 2–5 cm, which means it is some tenths of a wavelength wide. For this method, it is the efficient soil moisture sampling depth. The doubts introduced by disparities of surface roughness give rise to substantial issues in approximating moisture of soil from passive microwave assessment as determined by Wang (1983). The entailed spatial resolution is best met by active microwave (radar) sensors as recommended by Ulaby et al. (1986).

7.7. TECHNIQUES OF ACTIVE (RADAR) MICROWAVE

7.7.1. Synthetic Aperture Radar

A radar is fixed on a satellite, and microwave energy is produced on it. To calculate the brightness of microwave energy's strength produced inside a radar, which is dispersed or echoed from the Earth's surface or atmosphere, can be done on receipt behind at the satellite antenna. Dielectric constant of the substance of which the surface is made, orientation, surface roughness, and slope affect the volume of backscattered energy. Heavy rainfall weakens short; thus, it also relies upon rainfall interrupted due to the radar beam.

The SEASAT synthetic aperture sidelooking radar's (SAR) geometry is exhibited in Figure 7.14. (This sensor functioned at 1.27 GHz (23.50 cm wavelength) and manufactured a large aperture antenna by employing the movement of the satellite).

Whether or not a surface seems smooth or rough at a certain wavelength depends on the terrain feature's size concerning the electromagnetic wave's length as highlighted by Hall and Martinec (1985). The quality of being excellent specular reflectors is found in smooth water surfaces. All the microwave energy is speculated into space as they are not usually observed at right angles to the SAR. However, vertical, and horizontal surfaces may return a huge portion of the energy to the antenna as they may look like corner reflectors.

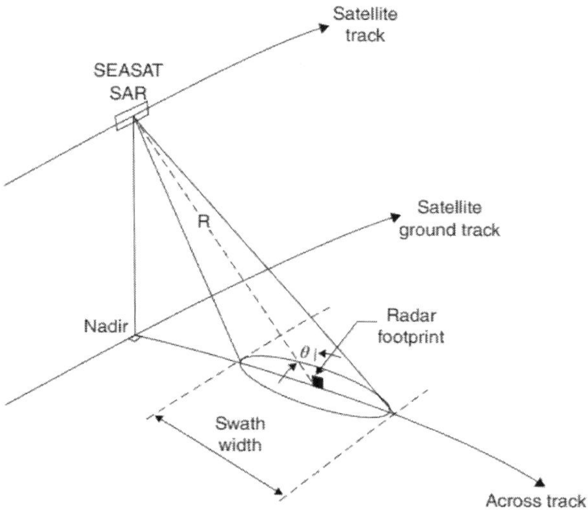

Figure 7.14. SEASAT SAR's geometry.

Source: *https://asf.alaska.edu/data-sets/sar-data-sets/seasat/seasat-about/.*

Given by (Fu and Holt, 1982) Figure 7.14 shows the figure resolution in the acrosstrack direction d_r.

$$d_r = \frac{c\tau}{2\sin\theta} \tag{13}$$

where; c is the light's speed; and τ is the effective pulse width. The picture area was 100 km × 4000 km while 25 m × 25 m was the greatest resolution attained by the SEASAT SAR. To determine the roughness of the oceans, a radar altimeter, functioning at 13.5 GHz (2.20 cm wavelength), was also carried by SEASAT.

The ability of the data for assessing soil moisture was identified quickly as SEASAT only worked for 100 days, unfortunately (Brisco et al., 1983; see also Ulaby et al., 1986; Ulaby et al., 1983).

Much of the SAR data was supplied through the launches of ERS1 and ERS2 functioning at 5.70 cm wavelength, both of which gave the utmost resolution of 125 m, and enabled the methods to be explored further. An appropriate representation in a catchment area of field size was provided by decreasing the resolution to 100 m as researchers proved that processing can possibly improve noise reduction. The return phase for a single satellite operation is 35 days which is the chief constraint on the value of the data. However, if one can access a data collection comprising of *in situ* soil moisture content calculations and backscatter coefficients, then describing a retrieval algorithm is a straightforward method.

At northwest England's Pennine moors, an algorithm was explained for recovering SMC; soil moisture content by a scientist. Displayed in Figure 7.15 are comparisons employing this method throughout the summer of 1996. The restrictions enforced by the satellite return interval are clear, although the satellite dimensions show a great degree of precision.

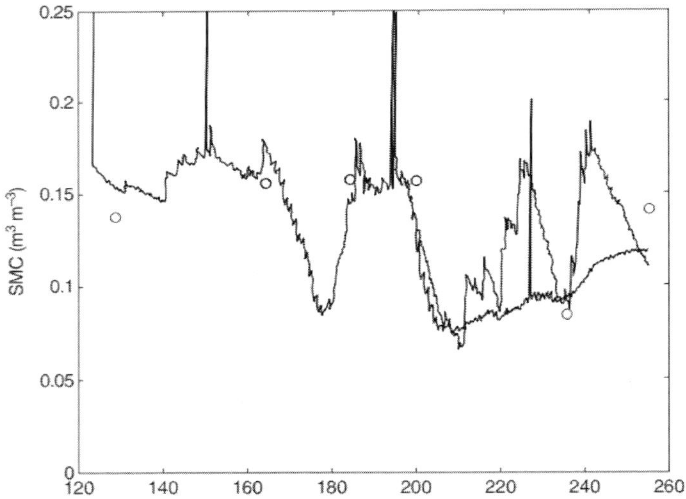

Figure 7.15. Measured by an ERS1 SAR (circles), capacitance probe (fainter line), and lysimeter (darker line), comparisons of soil moisture content are shown; instrumental noise is indicated through spikes.

Source: *https://www.booksamillion.com/search/search?type=author&query=Christopher+Collier+Collier&id=7520109285190.*

One can calculate the ET by a simple water balance equation as given below:

$$ET = R - P - (SMC_t - SMC_{t-1}) \quad (14)$$

where; P stands for the percolation out of the soil column; and R represents the rainfall in the time $t-1$ to t. To examine dynamics and tropical cyclone morphology, the pictures by the CSA, i.e., Canadian Space Agency RADARSAT 1/2 SAR have been utilized. In 2013–16, the RADARSAT Constellation Mission SAR and Sentinel1 functional quests will be launched by ESA and CSA (Li et al., 2013).

7.7.2. Radar Systems

Significant aptitude is presented by radar systems akin to radars which are groundbased. For instance, it is stated that undoubtedly, "given zero limitations on the antenna size, number of spacecraft and power, satellite active microwave systems can possibly offer precise worldwide divisions of rain rate." Utilizing adequately large antennas may help to conquer the issues of undesirable outcomes of radar echoes from the earth and radar beam filling. However, short wavelengths which are weakened by intense rainfall are needed for a practical antenna size. An area of view too expansive to settle important characteristics of precipitation will be given by the employment of nonweakening wavelengths (greater than 5.0 cm), along with the size constraints and antenna cost. Specifically, at offnadir angles, narrow beam swaths are formed by beam filling, reflectively gradients, and the return from the ground. At greater rainfall rates, the sensitivity is enhanced and the resolution is amplified by the use of a shorter attenuating wavelength. The attainable penetration into a storm is limited by the quick increase in attenuation at high and moderate rainfall rates. The issues are debated further by Kozu and Meneghini (1990).Utilizing two wavelengths, one attenuating and the other not is one probable solution put forth by a researcher. Either an extremely slow scan or a great radar pulse repetition frequency is indicated by the signal averaging time required to get the radar echo power's dependable estimate. Atlas et al. (1982) put forward the ensuing possibilities:

- Utilization of a minute wavelength inspecting radar functioning at 2 wavelengths (0.8 cm and 3.0 cm say). Through this, a practical size antenna will be enabled to get a narrow beamwidth. However, the accuracy which could be attained will be limited due to intense attenuation in rainfall of only reasonable intensity.
- Viewing the scattering characteristics of the surface as a calibration for reduction computation, the usage of a surface target attenuation radar. The issue to be overcome would be the scattering characteristics that are affected by soil moisture.

- Together with a passive microwave radiometer or with IR images, the employment of an altered radar altimeter functions at 13.50 GHz (2.20 cm wavelength). Regrettably, this would be limited to nadir measurements only (Table 7.2).

Table 7.2. TRMM PrecipitationAssociated Tools

Name of Instrument		**Precipitation Radar (PR)**	**LIS; Lightning Imaging Sensor**	**VIRS; Visible and Infrared Scanner**	**TMI; TRMM Microwave Imager**
Band frequencies and wavelengths		13.8 GHz	0.7774 μm	5 channels: 1.60 0.6, 3.7, 10.80, 12.0 μm	5 frequencies: 10.70, 19.40, 21.30, 37.0, 85.50 GHz
Spatial resolution (km)	Pre-boost	4.3 (vertical 250 m)	3.7	2.2	4.4 (at 85.5 GHz)
	Post-boost	5 (vertical 250 m)	4.3	2.4	5.1 (at 85.5 GHz)
Swath width (km)	Pre-boost	215	580	720	215
	Post-boost	247	668	833	247

Including the first satellite precipitation radar, the third option is the one that is approved for the NASA TRMM, whose launch took place in November 1997.

7.7.3. Tropical Rainfall Measuring Mission (TRMM)

In the areas of tropical and subtropical continents and oceans (40°S–40°N), the first thorough and complete data collection of fourdimensional divisions of latent heating and rainfall is given by TRMM which is a dual Japan-USA satellite mission. In TRMM, there are five instruments onboard. To determine the backscattered echo from cloud particles and precipitation, the PR, i.e., precipitation radar is employed which is an automatically scanning K_u band (13.80 GHz, 2.20 cm wavelength) radar. After amending the reflectivity calculations for attenuation (Adler et al., 2001), the rain retrievals are built on the reflectivity/rainrate (Z_e–R) relationships' application. A hybrid technique, i.e., the surface reference technique for which the comparative error reduces with growing path attenuation and the Hitschfeld-Bordan iterative attenuation correction method which seems to work well at low attenuations can be considered the basis of the reflectivities (Iguchi and Meneghini, 1994). The sampling rate is somewhat less than 11 hours over the United States (Simpson et al., 1996). The calibration steadiness is within 0.80 dB as indicated by scientists. Ground-based measurements were utilized for assessing ground radar calibration biases, and also for the precipitation measurements from TRMM. Displayed in Figure 7.16 is a sample of the figures of monthly rainfall calculated by TRMM (Anagnostou et al., 2001). The accessibility of services from TRMM and precipitation data are described by Liu et al. (2012).

7.8. THE SURFACE ENERGY BALANCE SYSTEM (SEBS)

The surface energy balance system (SEBS) has been described by Jong et al. (2008) in which many parameters, variables, and remote sensing techniques are employed.

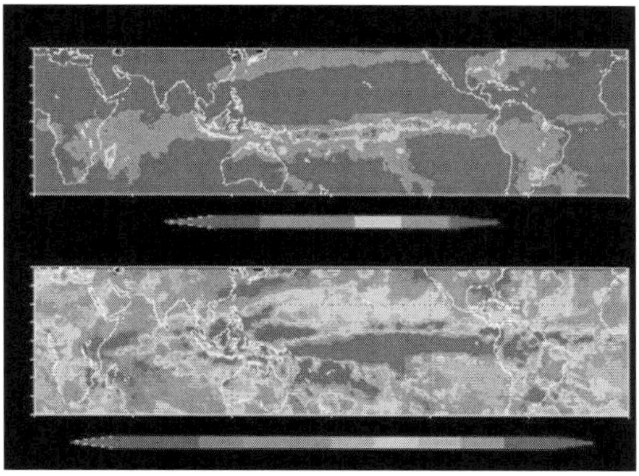

Figure 7.16. (a) January and February 1998 monthly mean rainfall by TRMM (mm); and (b) the distinction in 2month mean rainfall amidst 1998 and 1999 (NASA, GSFC).

Source: *https://www.researchgate.net/figure/The-panel-is-the-two-month-mean-rainfall-rate-mm-month-for-January-and-February-1998_fig1_252303161.*

Information regarding the facets of surface energy budgets is offered by thermal and optical sensors (Su, 2002). Soil moisture capacity and its spatial distribution are delivered by utilizing modeling. It is likely to combine satellite examinations to evaluate topsoil moisture amount by integrating the surface energy balance with the soil moisture balance. Estimates of leaf area index, topsoil moisture content, emissivity, height, the height of vegetation, albedo, surface temperature, and fractional vegetation cover are given by satellite examinations (see earlier in this chapter). Within a numerical model framework, these inspections are united with radiation data and meteorological measurements.

7.9. SATELLITE MEASUREMENT PROBLEM'S SUMMARY

Swift developments in gauging hydrometeorological limitations from space were attained over the last 40 years. As foretold by Collier (1985), through satelliteborne instrumentation, a lot of the input necessities to watershed replicas are possible to gratify, somewhat. Fully backedup functional measurement systems are needed for soil moisture, precipitation, snow cover, and ET as those parameters differ from day to day. It is vital to determine the cost-effectiveness and the relative merits of every method since numerous systems can offer measurements of similar parameters within an extensive portion of the electromagnetic spectrum. Two longer (19.0 GHz) and shorter (85.0 GHz) wavelengths were utilized to summarize the exchanges of passive microwaves with precipitation clouds (Figure 7.17).

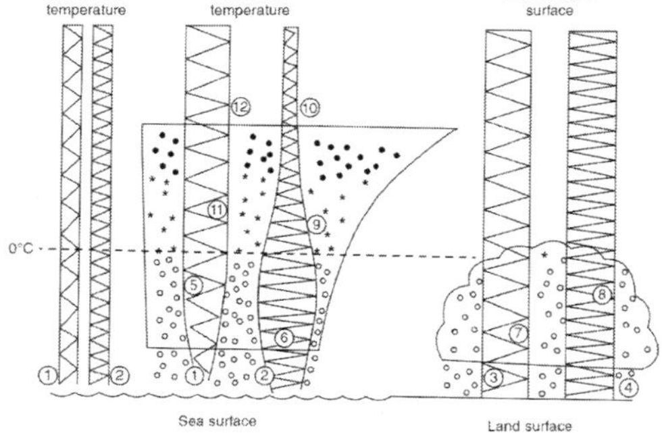

Figure 7.17. The communication between low (e.g., 19.0 GHz) and high (e.g., 85.0 GHz) frequency passive microwave with the surface and clouds of precipitation. The temperature and intensity of the upwelling radiation are signified by the width of the vertical column.

Note: In this picture, the drawn characteristics and related distinctions are: (A) the trivial sea surface's emissivity for high (2) and low (1) frequencies; (B) the great land surface's emissivity for both high (4) and low (3) frequencies; (C) raindrops' and cloud's emission, which saturates rapidly for the high frequency (6) but rises with vertically joined liquid water in the case of low frequency (5); (D) the indication at the low frequency of the water emissivity is disguised by the emissivity of the land surface (7); (E) the rain's saturated high-frequency emission (8) is not distinct from the background of the land surface (4); (F) ice precipitation particles began to backscatter down the high-frequency emission (9), producing temperatures of cold brightness (10), irrespective of the features of surface emission; (G) the ice does not check the low-frequency emission upwards (11), permitting its exposure as temperature of warm brightness above cloud top (12) (from Collier and Rosenfeld, 1999).

Source: https://www.bookdepository.com/author/Christopher-Collier.

As noted formerly, a great number of rain estimation techniques are devised by employing physical principles.

Assessments of the methods of measurement were conducted and the assessment of precipitation has had the greatest amount of effort put into it, mainly rainfall (e.g., Yumaz et al., 2005; Johnstone et al., 1985).

Employing the Griffith-Woodley method and rain gauge/radar data, the worst correlations amongst satellite approximate of rainfall were found for totals of half-hour rainfalls over expanses of 55.0 km^2 and the best correlations were attained over 9350.0 km^2 for 6-hour periods (the biggest area measured); as emphasized by the work of Augustine et al. (1981). Including an error of about 49% for rainfall totals over regions of 10^5 km^2 of 0.5 to 1 hour, the Austin bi-spectral and Lovejoy method offers the most precise approximations in all rainfall categories. Tuned to making evaluations in convective rainfall, other methods (Griffith and Woodley, 1981) may achieve this accuracy over regions of around 10^4 km^2. In the areas of 60% over 400 km^2, the accuracy of passive microwave methods is present. Monthly rainfall totals of 20% over 10^3 km^2 are provided by active microwave methods.

REFERENCES

1. Adler, R. F., Huffman, G. J., Bolvin, D. T., Curtis, S., & Nelkin, E. J., (2001). Tropical rainfall distributions determined using TRMM combined with other satellite and rain gauge information. *J. Appl. Meteorol.*, *39*, 2007–2023.
2. Adler, R. F., Huffman, G. J., Cheng, A., Ferraro, R., Xu, P. P., Janowtak, J., Rudolf, B., Schneider, U., Curtis, S., Bolvin, D., Gruber, A., Susskind, J., Arkin, P., & Nelkin, E., (2003). The version2 global precipitation climatology project (GPCP). Monthly precipitation analysis (1979–present). *J. Hydrometeorol.*, *4*, 1147–1167.
3. Anagnostou, E. N., Morales, C. A., & Dinku, T., (2001). The use of TRMM precipitation radar observations in determining ground radar calibration biases. *J. Atmos. Oceanic Technol.*, *18*, 616–628.
4. Atlas, D., Echweman, J., Meneghini, R., & Moore, R. K., (1982). The outlook for precipitation measurements from space. *Atmos. Ocean*, *20*, 50–61.
5. Augustine, J. A., Meitin, J. G., Griffith, C. G., & Woodley, W. L., (1981). An objective evaluation of the Griffith/Woodley rain estimation technique. In: *4th Conference on Hydrometeorology* (Vol. 4, pp. 134–140). Reno, NV. American Meteorological Society, Boston.
6. Barrett, E. C., & Grant, C. K., (1978). An appraisal of Landsat 2 cloud imagery and its implications for the design of a future meteorological observing system. *J. Brit. Interplanet. Soc.*, *31*, 3–10.
7. Barrett, E. C., & Martin, D. W., (1981). *The Use of Satellite Data in Rainfall Monitoring* (Vol. 1, pp. 10–29). Academic Press, London.
8. Barrett, E. C., (1970). The estimation of monthly rainfall from satellite data. *Mon. Weather Rev.*, *98*, 322–327.
9. Berg, W., Kummerow, C., Sapiano, M., RodriguezAlvarez, N., & Weng, F., (2012). A fundamental climate data record of microwave brightness temperature data from 25 years of SSM/I and SSMIS observations. *Gewex WCRP News* (Vol. 1, pp. 1–10).
10. Brisco, B., Ulaby, F. T., & Dobson, M. C., (1983). *Spaceborne SAR Data for LandCover Classification and Change Detection* (Vol. 83, pp. 10–44). Digest IEEE International Geoscience Remote Sensing Symposium (IGARSS'83), San Francisco.
11. Carlson, T. B., Dodd, J. K., Benjamin, S. G., & Cooper, J. N., (1981). Satellite estimation of the surface energy balance, moisture availability and thermal inertia. *J. Appl. Meteorol.*, *20*, 67–87.

12. Chang, T. C., Gloersen, P., Schmugge, T., Wilheit, T. T., & Zwally, H. J., (1976). Microwave emission from snow and glacier ice. *J. Glaciol.*, *17*, 23–39.
13. Collier, C. G., (1985). Remote sensing for hydrological forecasting. In: Rodda, J. C., (ed.), *Facets of Hydrology* (Vols. 1, II, pp. 1–24). Wiley, Chichester.
14. Davie, T., & Kelly, R., (1997). Spatial resolution and variation of soil moisture as detected by ERS1 SAR imagery. *Proceedings of the British Hydrological Society 6th National Hydrometeorology Symposium* (Vol. 6, pp. 5.11–5.17), Salford, Institute of Hydrology, Wallingford.
15. Delbeato, R., & Barrell, S. L., (1985). Rain estimation in extratropical cyclones using GMS imagery. *Mon. Weather Rev.*, *113*, 747–755.
16. Deleur, M. S., (1980). *Satellite Methods of Studying Snow Cover [in Russian]* (Vol. 1, pp. 10–29). Gridrometeoizdat, Leningrad.
17. Dey, B., Goswami, D. C., & Rango, A., (1983). Utilization of satellite snowcover observations for seasonal stream flow estimates in the Western Himalayas. *Nord. Hydrol.*, *3*, 257–266.
18. Dey, B., Gregory, A. F., & Moore, H., (1979). Application of satellite images for monitoring snowline in the Yukon and North West territories. *Polar Record*, *19*, 145–154.
19. Dittberner, G. J., & Vonder Haar, T. H., (1973). Large scale precipitation estimates using satellite data: Application to the Indian Monsoon. *Arch. Met. Geophys. Biokl. B*, *21*, 317–334.
20. Doneaud, A. A., Miller, J. R., Johnson, L. R., Vonder, H. T. H., & Laybe, P., (1985). The area timeintegral technique to estimate convective rain volume over areas applied to satellite data: A preliminary investigation. *Preprints 6th Conference on Hydrometeorology* (Vol. 1, pp. 238–245). American Meteorological Society, Boston.
21. Follansbee, W. A., & Oliver, V. J., (1975). A comparison of infrared imagery and video pictures in the estimation of daily rainfall from satellite data. *NOAA Technical Memorandum NESS 62* (pp. 1–20), Washington, DC.
22. Follansbee, W. A., (1973). Estimation of average daily rainfall from satellite cloud photographs. *NOAA Technical Memorandum NESS 44* (pp. 10–29), Washington, DC.
23. Follansbee, W. A., (1976). Estimation of daily precipitation over China and the USSR using satellite imagery. *NOAA Technical Memorandum NESS 81* (pp. 10–29), Washington, DC.

24. Foster, J. L., (1983). Nighttime observations of snow using visible imagery. *Int. J. Remote Sens., 4*, 785–791.

25. Foster, J. L., Rango, A., Hall, D. K., Chang, A. T. C., Allison, L. J., & Diesen, B. C. III, (1980). Snowpack monitoring in North America and Eurasia using passive microwave satellite data. *Remote Sens. Environ., 10*, 205–298.

26. Fox, N. I., Saich, P., & Collier, C. G., (2000). Estimating the surface water and radiation balance in an upland area from space. *Int. J. Remote Sens., 21*, 2985–3002.

27. Fu, L., & Holt, B., (1982). Seasat views oceans and sea ice with synthetic aperture radar. *JPL 81–120* (pp. 1–30). NASA Jet Propulsion Laboratory, Pasadena, CA.

28. Gloersen, P., & Salomonson, V. V., (1975). Satellites: new global observing techniques for ice and snow. *J. Glaciol., 15*, 373–389.

29. Griffith, C. G., & Woodley, W. L., (1981). The estimation of convective precipitation from GOES imagery with the Griffith/Woodley technique. In: Atlas, D., & Thiele, O. W., (eds.), *Precipitation Measurements from Space Workshop Report* (Vol. 1, pp. D154–D158). NASA Goddard Space Flight Center.

30. Griffith, C. G., Woodley, W. L., Browner, S., Teijeiro, J., Maier, M., Martin, D. W., Stout, J., & Sikdar, D. N., (1976). Rainfall estimation from geosynchronous satellite imagery during the daylight hours. *NOAA Technical Report ERL 356WPM07* (pp. 1–20), Boulder, CO.

31. Griffith, C. G., Woodley, W. L., Grube, P. G., Martin, D. W., Stout, J., & Sikdar, D. N., (1978). Rain estimation from geosynchronous satellite imagery: Visible and infrared studies. *Mon. Weather Rev.*, 106, 1153–1171.

32. Gruber, A., (1973). Estimating rainfall in regions of active convection. *J. Appl. Meteorol., 12*, 110–118.

33. Gurney, R. J., & Camillo, P. J., (1984). Modelling daily evapotranspiration using remotely sensed data. *J. Hydrol., 69*, 305–324.

34. Hall, D. K., & Martinec, J., (1985). *Remote Sensing of Ice and Snow* (Vol. 1, pp. 5–29). Chapman and Hall, London.

35. Harries, J. E., (1994). *Earthwatch: The Climate from Space* (Vol. 1, pp. 5–19). Wiley, Chichester.

36. Harris, R., (1987). *Satellite Remote Sensing* (Vol. 1, pp. 5–16). Routledge & Kegan Paul, London.

37. Hsu, K., Gao, X., Sorooshian, S., & Gupta, H. V., (1997). Precipitation estimation from remotely sensed information using artificial neural networks. *J. Appl. Meteorol.*, *36*, 1176–1190.
38. Huffman, G. J., et al., (1997). The global precipitation climatology project (GPCP). Combined precipitation datasets. *Bull. Am. Meteorol. Soc.*, *78*, 36–50.
39. Idso, S. B., Jackson, R. D., & Reginato, R. J., (1975a). Detection of soil moisture by remote surveillance. *Am. Sci.*, *63*, 549–557.
40. Idso, S. B., Schmugge, T. J., Jackson, R. D., & Reginato, R. J., (1975b). The utility of surface temperature measurements for the remote sensing of soil water status. *J. Geophys. Res.*, *80*, 3044–3049.
41. Iguchi, T., & Meneghini, R., (1994). Intercomparison of single frequency methods for retrieving a vertical rain profile from airborne or spaceborne radar data. *J. Atmos. Oceanic Technol.*, *11*, 1507–1516.
42. Iguchi, T., Meneghini, R., Awaka, J., Kozu, T., & Okamoto, K., (2000). Rain profiling algorithm for TRMM precipitation radar data. *Remote Sens. Appl. Earth Atmos. Ocean*, *25*, 973–976.
43. Inomata, H., Okamoto, K., Ajima, T., Masuko, H., Yohidada, S., & Gugono, M., (1981). Remote sensing of rainfall rates using airborne microwave rainscatterometer/radiometer. *Proceedings 15th International Symposium on Remote Sensing of Environment* (Vol. 15, pp. 1–29). Ann Arbor, MI, USA.
44. Jackson, R. D., Schmugge, T. J., Nicks, A. D., Coleman, G. A., & Engman, E. T., (1981). Soil moisture updating and microwave remote sensing for hydrological simulation. *Hydrol. Sci. Bull.*, *26*, 39/1981, 305–319.
45. Johnstone, K. J., Hogg, W. D., Hanssen, A. J., Niitsoo, A., & Polavarapu, V. L., (1985). An evaluation of some satellite rainfall estimation techniques. *Preprints 6th Conference on Hydrometeorology* (Vol. 1, pp. 233–237). American Meteorological Society, Boston.
46. Jong, S. De, Kwast, H. Van Der, Addink, E., & Su, B., (2008). Remote sensing for hydrological studies. Chapter 15. In: *Climate and the Hydrological Cycle* (pp. 297– 320). IAHS Special Publication 8, IAHS Press, CEH Wallingford.
47. Klaassen, W., & Van Den Berg, W., (1985). Evapotranspiration derived from satellite observed surface temperatures. *J. Clim. Appl. Meteorol.*, *24*, 412–424.

48. Kozu, T., et al., (2001). Development of precipitation radar onboard the tropical rainfall measuring mission (TRMM). *IEEE Trans. Geosci. Remote Sens.*, *39*, 102–116.
49. Kunzi, K. F., Patil, S., & Rott, H., (1982). Snow cover parameters retrieved from Nimbus7 scanning multichannel microwave radiometer (SMMR). data. *IEEE Trans. Geosci. Remote Sens.*, *GE20*, 452–467.
50. Lethbridge, M., (1967). Precipitation probability and satellite radiation data. *Mon. Weather Rev.*, *95*, 487–490.
51. Li, X., Zhang, J. A., Yang, X., Pichel, W. G., DeMaria, M., Long, D., & Li, Z., (2013). Tropical cyclone morphology from spaceborne synthetic aperture radar. *Bull. Am. Meteorol. Soc.*, *94*, 215–230.
52. Liljas, E., (1982). Automated techniques for the analysis of satellite cloud imagery. In: Browning, K. A., (ed.), *Nowcasting* (Vol. 1, pp. 167–176). Academic Press, London.
53. Liu, Z., Ostrenga, D., Teng, W., & Kempler, S., (2012). Tropical rainfall measuring mission (TRMM). Precipitation and services for research and applications. *Bull. Am. Meteorol. Soc.*, *93*, 1317–1325.
54. Lovejoy, S., & Austin, G. L., (1979a). The delineation of rain areas from visible and IR satellite data for GATE and midlatitudes. *Atmos. Ocean*, *17*, 77–92.
55. Lovejoy, S., & Austin, G. L., (1979b). The sources of error in rain amount estimating schemes for GOES visible and IR satellite data. *Mon. Weather Rev.*, *107*, 1048–1054.
56. Lovejoy, S., & Austin, G. L., (1980). The estimation of rain from satelliteborne radiometers. *Q. J. R. Meteorol. Soc.*, *106*, 255–276.
57. Martin, D. W., & Howland, M. R., (1986). Grid history: A geostationary satellite technique for estimating daily rainfall in the tropics. *J. Clim. Appl. Meteorol.*, *25*, 184–195.
58. Martin, M. J., Hines, A., & Bell, M. J., (2007). Data assimilation in the FOAM operational shortrange ocean forecasting system: A description of the scheme and its impact. *Q. J. R. Meteorol. Soc.*, *133*, 981–995.
59. Martinec, J., & Rango, A., (1981). Areal distribution of snow water equivalent evaluated by snow cover monitoring. *Water Resour. Res.*, *17*, 1480–1488.
60. McElroy, J. H., & Schneider, S. R., (1985). *The Space Station Polar Platform: Integrating Research and Operational Missions* (pp. 1–29). NOAA Technical Report NESDIS 19, US Dept. of Commerce, Washington, DC.

61. McGinnis, D. F. Jr, Scofield, R. A., Schneider, S. R., & Berg, C. P., (1979). *Satellites as an Aid to Water Resources Managers* (pp. 1–33). American Society of Civil Engineers Convention, Boston, MA, preprint 3486.

62. Meneghini, R., & Kozu, T., (1990). *Spaceborne Weather Radar*. Artech, Boston.

63. Negri, A. J., & Adler, R. F., (1981). Relation of satellitebased thunderstorm intensity to radar estimated rainfall. *J. Appl. Meteorol.*, *20*, 288–300.

64. Negri, A. J., Adler, R. F., & Wetzel, P. J., (1984). Rain estimation from satellites: An examination of the Griffith Woodley technique. *J. Clim. Appl. Meteorol.*, *23*, 102–116.

65. O'Brien, H. W., & Munis, R. H., (1975). Red and nearinfrared spectral reflectance of snow. In: Rango, A., (ed.), *Operational Applications of Satellite Snowcover Observations* (pp. 345–360). Proceedings Workshop, South Lake, Tahoe, CA, NASA SP391, NASA, Washington, DC.

66. Odegaard, H. A., Anderson, T., & Ostrem, G., (1980). *Applications of Satellite Data for Snow Mapping in Norway* (pp. 92–106). NASA CP 2116.

67. Ramamoorthi, A. S., & Rao, P. S., (1985). Inundation mapping of the Sahibi river flood of 1977. *Int. J. Remote Sens.*, *6*(3, 4), 443–445.

68. Rango, A., & Itten, K. I., (1976). Satellite potentials in snow cover monitoring and runoff prediction. *Nord. Hydrol.*, *7*, 209–230.

69. Rango, A., & Martinec, J., (1979). Application of a snowmeltrunoff model using Landsat data. *Nord. Hydrol.*, *10*, 225–238.

70. Rango, A., & Peterson, R., (1980). *Operational Application of Satellite Snowcover Observations* (Vol. 2116, pp. 1–22). NASA Conference Publication.

71. Rango, A., Chang, A. T. C., & Foster, J. L., (1979). The utilization of spaceborne microwave radiometers for monitoring snowpack properties. *Nord. Hydrol.*, *10*, 25–40.

72. Rango, A., Salomonson, V. V., & Foster, J. L., (1977). Seasonal streamflow estimation in the Himalayan region employing meteorological satellite snow cover observations. *Water Resour. Res.*, *13*, 109–112.

73. Rees, W. G., (1993). *Physical Principles of Remote Sensing* (Vol. 1, pp. 5–19). Cambridge University Press, Cambridge.
74. Reynolds, D. W., Vonder Haar, T. H., & Grant, L. O., (1978). Meteorological satellites in support of weather modification. *Bull. Am. Meteorol. Soc.*, *59*, 269–281.
75. Rosenfeld, D., & Collier, C. G., (1999). Estimating surface precipitation. Chapter 4.1. In: Browning, K. A., & Gurney, R. J., (eds.), *Global Energy and Water Cycles* (Vol. 1, pp. 124–133). Cambridge University Press, Cambridge.
76. Salman, A. A. B., (1983). Using Landsat imagery interpretation for underground water prospection around Qena Province, *Egypt. Int. J. Remote Sens.*, *4*, 179–189.
77. Salomonson, V. V., & Hall, D. K., (1979). A review of LandsatD and other advanced systems relative to improving the utility of space data in water resources management. In: Rango, A., & Peterson, R., (eds.), *Operational Applications of Satellite Snowcover Observations* (pp. 281–296). Proceedings Workshop, Sparks, NV, NASA CP2116, NASA, Washington, DC.
78. Schanda, E., (1986). *Physical Fundamentals of Remote Sensing* (Vol. 1, pp. 10–30). Springer, Berlin.
79. Scofield, R. A., & Oliver, V. J., (1977a). A scheme for estimating convective rainfall from satellite imagery. *NOAA Technical Memorandum NESS 86* (pp. 10–29). Washington, DC.
80. Scofield, R. A., & Oliver, V. J., (1977b). Using satellite imagery to estimate rainfall from two types of convective systems. In: *Preprints 11th Technical Conference on Hurricanes and Tropical Meteorology* (Vol. 1, pp. 204–211). Miami Beach, FL, American Meteorological Society, Boston.
81. Scofield, R. A., (1981). Analysis of rainfall from flash flood producing thunderstorms using GOES data. *Proceedings IAMAP Symposium, Hamburg, ESA SP175* (pp. 51–58).
82. Scofield, R. A., (1982). A satellite technique for estimating rainfall from flash flood producing thunderstorms. In: Johnson, A. I., & Clark, R. A., (eds.), *Proceedings International Symposium on Hydrometeorology* (Vol. 1, pp. 121–120). Denver, CO, American Water Resources Association, Bethesda, MD.

83. Scofield, R. A., (1984). Satellitebased estimates of heavy precipitation. *Recent Advances in Civil Space Remote Sensing, SPIE 481*, pp. 84–91.
84. Simpson, J. C., Kummerow, C., Tao, W. K., & Adler, R. F., (1996). On the tropical rainfall measuring mission (TRMM). *Meteorol. Atmos. Phys.*, *60*, 19–36.
85. Sorooshian, S., Hsu, K., Gao, X., Gupta, H. V., Iman, B., & Braithwaite, D., (2000). Evolution of PERSIANN system satellitebased estimates of tropical rainfall. *Bull. Am. Meteorol. Soc.*, *81*, 2035–2046.
86. Spencer, R. W., (1986). A satellite passive 37GHz scatteringbased method for measuring oceanic rain rates. *J. Clim. Appl. Meteorol.*, *25*, 754–766.
87. Spencer, R. W., Hinton, B. B., & Olson, W. S., (1983). Nimbus7 37 GHz radiances correlated with radar rainrates over the Gulf of Mexico. *J. Clim. Appl. Meteorol.*, *22*, 2095–2099.
88. Stout, J. E., Martin, D. W., & Sikdar, D. N., (1979). Estimating GATE rainfall with geosynchronous satellite images. *Mon. Weather Rev.*, *107*, 505–598.
89. Su, Z., (2002). The surface energy Balance (SEBS). For estimation of turbulent heat fluxes. *Hydrol. Earth Syst. Sci.*, *6*, 85–99.
90. Theon, J., (1973). A multispectral view of the Gulf of Mexico from Nimbus5. *Bull. Am. Meteorol. Soc.*, *54*, 934–937.
91. Tsonis, A. A., & Isaac, G. A., (1985). On a new approach for instantaneous rain area delineation in the midlatitudes using GOES data. *J. Clim. Appl. Meteorol.*, *24*, 1208–1218.
92. Ulaby, F. T., Dobson, M. C., & Brunfeldt, D. R., (1983). Improvements of moisture estimation accuracy of vegetationcovered soil by combined active/passive microwave remote sensing. *IEEE Trans. Geosci. Remote Sens.*, *GE21*, 300–307.
93. Ulaby, F. T., Moore, R. K., & Fung, A. K., (1986). *Microwave Remote Sensing: Active and Passive*: *From Theory to Applications* (Vol. III, pp. 1–33). Artech, Boston.
94. Van de Griend, A. A., & Engman, E. T., (1985). Partial area hydrology and remote sensing. *J. Hydrol.*, *81*, 211–251.
95. Wang, J. R., (1983). Passive microwave sensing of soil moisture content: The effects of soil bulk density and surface roughness. *Remote Sens. Environ.*, *13*, 329–344.

96. Wang, J., & Wolff, D. B., (2012). Evaluation of TRMM rain estimates using ground measurements over Central Florida. *J. Appl. Meteorol. Climatol.*, *51*, 926–940.
97. Wetzel, P. J., Atlas, D., & Woodward, R. H., (1984). Determining soil moisture from geosynchronous satellite infrared data: A feasibility study. *J. Clim. Appl. Meteorol.*, *23*, 375–391.
98. Wilheit, T. H., (1975). The electrically scanning microwave radiometer (ESMR) experiment. In: *Nimbus6 User Guide* (pp. 87–108). NASA Goddard Space Flight Center.
99. Wilheit, T. H., Chang, A. T. C., Rao, M. S. V., Rodgers, E. B., & Theon, J. S., (1977). A satellite technique for quantitatively mapping rainfall rates over the oceans. *J. Appl. Meteorol.*, *16*, 551–560.
100. Wilheit, T. H., Theon, J., Shenk, W., & Allison, L., (1973). Meteorological interpolation of the images from Nimbus5 electrically scanned microwave radiometer. *NASA X65173189* (pp. 1–22), Goddard Space Flight Center.
101. Woodley, W. L., Griffith, C. G., Griffin, J. S., & Stromatt, S. C., (1980). The inference of GATE convective rainfall from SMS1 imagery. *J. Appl. Meteorol.*, *19*, 388–408.
102. Wu, R., Weinman, J. A., & Chin, R. T., (1985). Determination of rainfall rates from GOES satellite images by pattern recognition technique. *J. Atmos. Oceanic Technol.*, *2*, 314–330.
103. Wylie, D. P., & Laitsch, D., (1983). The impacts of different satellite data on rain estimation schemes. *J. Appl. Meteorol.*, *22*, 1270–1281.
104. Wylie, D. P., (1979). An application of a geostationary satellite rain estimation technique to an extratropical area. *J. Appl. Meteorol.*, *18*, 1640–1648.
105. Yumaz, K. K., Hogue, T. S., Hsu, K.L., Sorooshian, S., Gupta, H. V., & Wagener, T., (2005). Intercomparison of rain gauge, radar and satellitebased precipitation estimates. *J. Hydrometeorol.*, *6*, 497–517.
106. Zwally, H. J., & Gloersen, P., (1977). Passive microwave images of the polar regions and research applications. *Polar Record*, *18*, 431–450.

Chapter 8

Hydrometeorology in the Urban Environment

CONTENTS

8.1. INTRODUCTION

According to United Nations (UN, 2001), in the previous year 2000, 47% of the world's population lived in cities, and by 2030, that number is expected to increase to 60%. In the developed countries, at least 75% of the population inhabited cities, with that number expected to increase to 83% by the year 2030. In the world's less developed countries, 18% of the population resided in cities in 1950, and this is projected to increase to 56% by the year 2030.

In the year 2000, only 3.7% of the world's population lived in urban areas with populations of 10 million or even more, while 24.8% continued to live in cities with populations of less than 500,000. From 40 in 2001 to 58 in 2005, the number of cities with populations exceeding 5 million people increased. After all, as the number of mega-cities has grown, so has the number of smaller urban areas, and the implications of large cities particularly on hydrology and weather can be very crucial (Collier, 2006).

8.2. THE URBAN BOUNDARY LAYER AND THE CYCLE OF WATER

Energy fluxes and airflow are influenced by urban morphology, resulting in phenomena like the heat island of the urban region and convective rainfall commencement (see for instance Collier, 2006). The UBL (urban boundary layer) is a transition zone between the earth and the surrounding formed by longitudinal and transverse flows across an urban area. The surface of most metropolitan centers is quite diverse, with substantial differences resulting from construction types like business quarters, high rise, etc., and external land uses such as rivers, parks, etc. This has a big impact on how we perceive weather patterns and measured energy budgets. The UHI (urban heat island) effect occurs when the temperature in the core of cities differs significantly from the surrounding countryside (see for instance Oke, 1987). Figure 8.1 illustrates how the cloud base at the pinnacle of the boundary layer varies all across the rural-urban divide.

The majority of cities produce heat as a result of human activity. Large expanses of concrete and asphalt with heat capacities and high albedos allow incoming radiation to be converted and stored as sensible heat more efficiently than land in rural expanses.

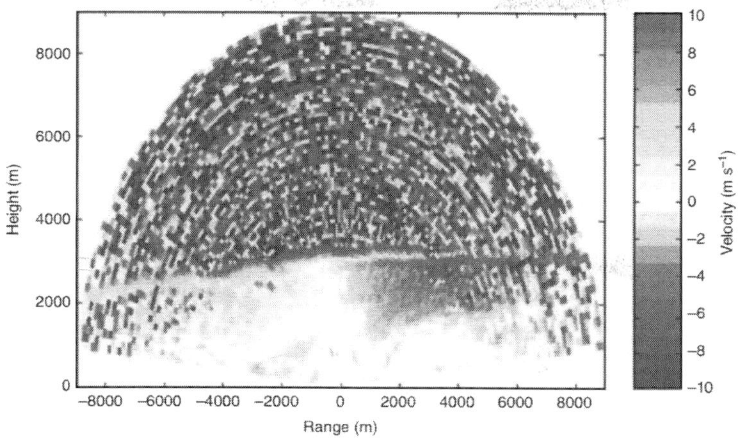

Figure 8.1. A cross-section (vertical) of the rural-urban boundary in London (west), exhibiting cloud base fluctuation at the boundary layer top as recorded by Doppler lidar.

Note: The image on the left is above the area of west side rural region, and the one on the right is over the metropolitan area to the eastern side. The colors show the Doppler radial velocities strength in m s – 1 in the direction of (positive) and far to (negative) the lidar in m s – 1. The colors just above boundary layer indicate noise in the findings.

Source: https://onlinelibrary.wiley.com/doi/book/10.1002/9781118414965.

As a result, as previously mentioned, the lower parts of the UBL are frequently warmer than their surrounding rural areas (Oke, 1982; Chandler, 1976). The impact of building configuration on the radiation terms of the surface energy balance has been shown by Harmon et al. (2004) and others to be the main reason for surface temperature divergences between cities and suburbs. Throughout many cities, UHIs have been discovered.

While the power of the UHI is directly proportional to urban metropolitan population (Oke, 1987), research from Lodz, Poland (Klysik and Fortuniak, 1999) suggests that the size of the population can be a substitute for the repercussions of structure development density and the onset of various forms of artificial surfaces. As Arnfield (2003) points out, during the 20 years(last) or so, it has made clear that there are numerous kinds of UHIs, each with its own set of features and regulated by its own set of interrelating energy exchange processes. The UHI-induced mesoscale confluence-that can dramatically change the layer of boundary- and its connection with some other local publications is one of the prominent methods that can demonstrate the noticeable urban impacts on the rainy season during the summer (see for instance Cotton and Pielke, 2007; Shepherd and Burian, 2003).

Figure 8.2 depicts the effect of cover(impervious) on the cycle of water. With natural ground cover, 25% of rain penetrates the groundwater and thus only 10% runs off. As the imperviousness of a surface rises, very little water penetrates, and much more runs off. Over 50% of all rain in densely populated areas becomes surface runoff.

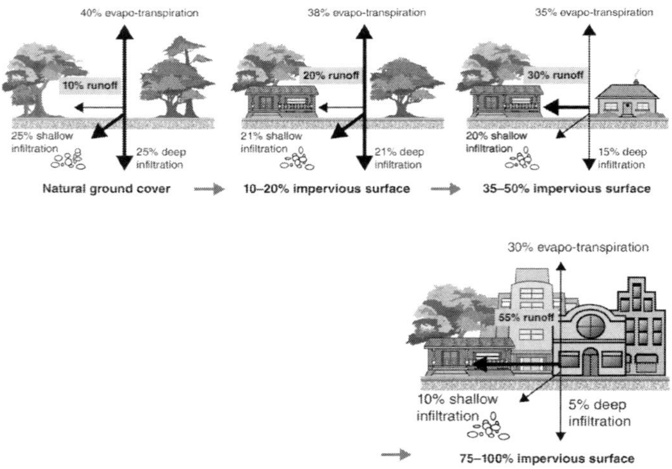

Figure 8.2. The impact of impervious cover on the water cycle.

Source: www.coastal.ca.gov/nps/watercyclefacts.pdf.

8.3. URBAN RAINFALL AND DEVELOPMENT

Urbanization, according to Shepherd and Burian (2005), changes the natural landscape's texture and disrupts Earth's processes system. The water cycle is altered during building projects because the soil layer is altered, vegetation is removed, and structures built and infrastructure of drainage are initiated. Normally, infrastructural projects within a watershed decrease penetration and aquifer recharge, enhance surface runoff rates and volumes, minimize soil moisture, and change the spatial patterns and quantity of water and energy fluxes and storage.

Rainfall is influenced by urban areas, as has been known for decades. Changnon (1981) summarized Project METROMEX (Metropolitan Meteorological Experiment), which showed a 30% to 40% enhance in rainfall at La Porte, USA, Indiana, situated Chicago-Gary downwind area, due to urban expansion during 1925–70 (Changnon and Huff, 1972). Analogous effects have been reported somewhere else, including in Israel (Goldreich, 1987). Huff and Changnon (1972) discovered an afternoon all-out of rainfall 25 kilometers east of Missouri, St Louis, and a second maximum 30 kilometers north of the city within 2100–2400 Central Standard Time in the overall rainfall of summer illustrated in Figure 8.3 (Huff and Changnon, 1986).

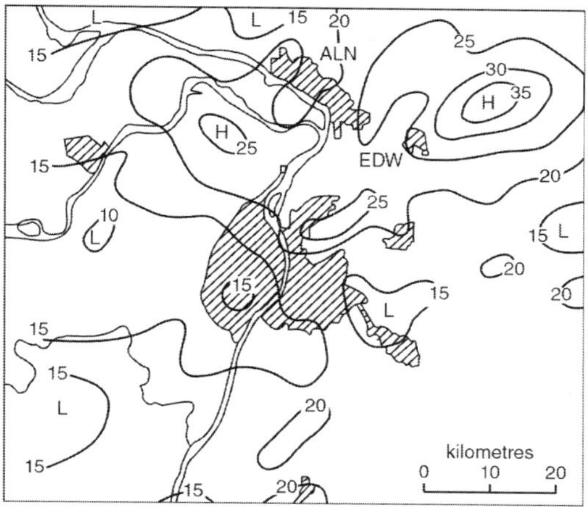

Figure 8.3. Between 2100 and 2400 CST in 1971–1975, cumulative summer rainfall within centimeters near St. Louis, Missouri, USA. Major urban locations are depicted in the shaded region (also labelled EDW, ALN). L and H represent low and high (compared to average) rainfall areas, respectively.

Source: https://onlinelibrary.wiley.com/doi/book/10.1002/9781118414965.

According to new research, the increased frequency of rain starts over towns and cities is linked to three factors to be considered by Changnon et al. (1976):

- Confluence zones in which clouds form due to mechanical and hydrodynamic impacts;
- The presence of gargantuan nuclei speeds up the coalescence method; and
- Extra clouds and significantly larger include instability as a result of thermodynamic ramifications.

Carraca and Collier (2007) investigated the relative significance of surface morphology distribution and category, as well as urban heating. The implications of urban heterogeneity were investigated using a basic model of surface sensible heat flux. To check the model's authenticity, sensitivity tests were conducted, and tests with a diagram of urban morphology were utilize to analyze the influence of various aspects of construction arrays. It was discovered that elevated buildings with a small footprint can have just as much effect as lower-rise buildings with a larger footprint. The urban environment causes significant variations in surface sensible heat flux. Convective cells can be triggered by the heat flux(sensible) contribution caused by elevated buildings in the city middle when the atmospheric layer of boundary is unreliable, according to information from a C-band radar positioned in the northern part of Greater Manchester (United Kingdom).

In comparison to surrounding areas, Shepherd, and Burian (2005) observed that the mean yearly and hot weather diurnal rainfall pattern in the Houston (1984–1999) has an 18 to 28% high fraction of the daily average rainfall taking place between noon and 4 p.m. This improvement could be related to the UHI (Burian and Shepherd, 2003). Kishtawal et al. (2009) looked into a potential connection between urbanization trends identified by remotely sensed data and precipitation patterns in the Indian summer monsoon area. According to their findings, there is sufficient statistical reason to prove that the reported growing trend in the intensity of extreme rainfall occurrence over the Indian monsoon geographical area is highly prone to be over areas where land use alters at a rapid pace due to urbanization. Furthermore, satellite-based rainfall estimations show that urban regions are highly(less) prone to undergo heavier (lighter) rates of precipitation than non-urban regions.

Lin et al. (2011) used the next-generation framework on mesoscale level, the WRF (denote the Weather Research and Forecasting), in conjunction with the model of Noah land surface and the urban canopy model (UCM), to assess the effects of the effect of UHI on precipitation over a complex spatial environment in northern Taiwan. As relative to 25-category land use categorization, the initial US-GS model greatly enhanced results of simulation for the pattern of accumulation rainfall that leads to better land use classification sourced from MODIS (modest resolution imaging spectroradiometer) satellite data (the MODIS case). In the urban case, the precipitation system developed later but was better than in the non-urban case. In comparison to radar observations, simulation results were fairly accurate. In the MODIS case, the rainfall system was bolstered not only downwind of the city over the mountainous region, but also upwind of the city over the plain area. According to the simulation results, proper land use categorization is critical for an urban heat island model-based research. The UHI effect has a significant impact on thermal and dynamic processes. It has an impact on the site of thunderstorms and precipitation in northern Taiwan's complex spatial environment.

8.4. SEWER FLOODING

There are many forms of inundation in urban areas, including flooding caused by natural disasters and infrastructure failures, pluvial flooding, groundwater flooding, fluvial flooding, and flooding caused by asset failures.

Sewer flooding occurs when wastewater exits from a pipe via a pit or drain, or backs up through toilets, sinks, and baths. Sewer flooding can result from a sewer pipe blockage, too much water entering the sewers from storm runoff-from fields and roads- and teeming waterways and rivers, equipment failure, or the sewer being too small to handle the amount of sewage penetrating it. The source of the problem could be a long way from where the flooding is occurring. Internal flooding occurs when sewage leaks into a building. External flooding occurs when water floods orchards or nearby regions such as roads and public spaces. Figure 8.4 illustrates the differences between consolidated and segregated drainage systems. Many new developments in ancient cities have coupled sewers, which transfer rainwater and sewage together.

In some places, combined sewer flows are at or above potential, while in others, they are below potential. As a result, in-system storage is used to allow flood spikes to pass through the system without triggering sewers to overflow or wastewater to be drained into natural waterways. In-pipes, offline retention tanks, and in-line storage tanks may all be used to store this information.

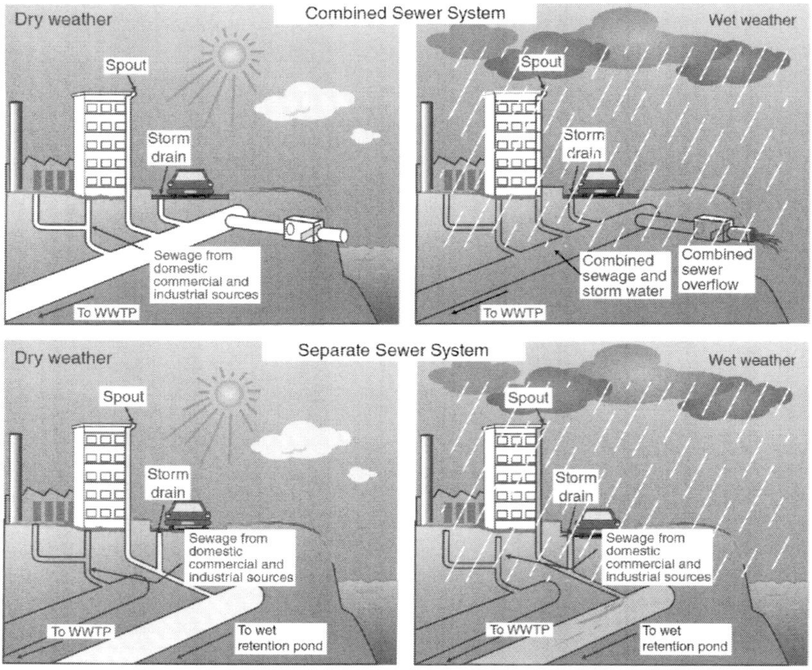

Figure 8.4. Wastewater treatment plants (WWTPs) are one type of sewage system.

Source: https://onlinelibrary.wiley.com/doi/pdf/10.1002/9781118414965.ch14.

The actual control operation strategy necessitates the authority to influence and direct flows in the sewer system via gates, diversion regulators, and dams. This technique necessitates detailed system models as well as algorithms that direct control activities. Direct control, as Adams and Papa (2000) figure out, does not reduce the level of combined sewage in a system; rather, it alters the spatial and temporal variability of that volume at the sewer outfall. This has the potential to reduce combined sewer overflows significantly (see for instance Labadie et al., 1975).

Surcharged and inundated sewer systems necessitate simulation models to adequately represent the hydraulic phenomena. As shown in Figure 8.5, 'dual drainage modeling' refers to the discrete surface flow and its interplay with sewer flow. This consists of:

- Areas connected entirely by closed drains, such as roofs;
- Areas where there is a lot of surface flow;

- Areas that are connected by surface inlets and closed drains, such as parking lots;
- Areas that do not have access to the sewer system;
- Areas on private property that drain to the street;
- Areas that are not prone to flooding.

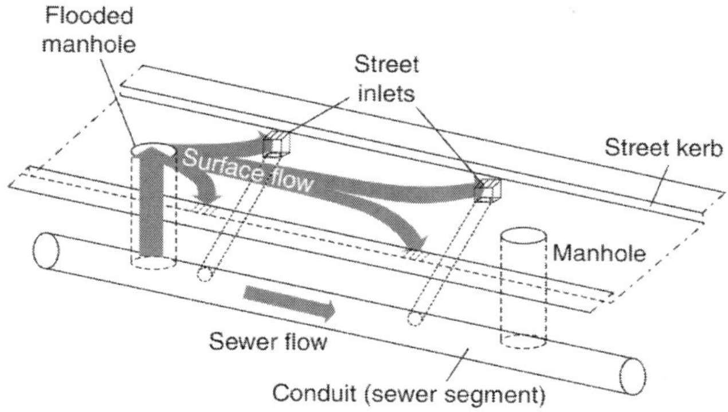

Figure 8.5. Sewer and surface flow interactions (dual drainage concept).

Source: https://www.researchgate.net/figure/Interaction-of-surface-and-sewer-flow-dual-drainage-concept-Schmitt-et-al-2004_fig1_266342445.

Hydraulic surface flow modeling focused on the Navier-Stokes equations is described by Schmitt et al. (2004). According to Deletic et al. (2012), the current state of knowledge concerning complexities in urban drainage designs is poor. The improvement in modeling surface runoff in metropolitan areas is discussed in further sections.

8.5. URBAN AREAS SURFACE RUNOFF

Flooding of surface water can be induced or exacerbated in urban regions in which the ground is largely made up of hard surfaces like tarmac or concrete, allowing precipitation to run off instead of seep further into soil. Figure 8.6 depicts the influence of urban expansion on Thompson Creek in California's Santa Clara Valley. The key difference among post-and pre-development situations as the development progressed was the much higher volume of runoff produced just after improvement (California Water and Land Use Partnership, WALUP). Severe weather events are becoming more often in the UK and abroad, presumably as a result of climate change, compounding the dangers for houses in small urban areas that are regarded to be at risk of pluvial flooding. According to the Environment Agency, roughly 3.8 million homes in England are vulnerable to flooding of surface water, comparable to only 2.4 million homes at danger of coastal or fluvial flooding. Identical issues emerge in other parts of the world where surface runoff has a variety of elements (see for example Choe et al., 2002).

As there are no such clearly understood floodplains such as for seas and rivers so accurately forecasting pluvial flood threats is challenging. The ability of infrastructure like sewers, buildings, and street furniture to withstand heavy rain must all be addressed (see for example Schellart et al., 2011). Hull was one of several cities and towns throughout England to be hit in the hot weather of 2007 after bursts of severe rainfall created several flooding incidents, resulting in insured flood damages totaling £3 billion throughout the country-the biggest sum previously paid by British insurers. Due to a convergence of global warming and a predicted rise in the UK's urban inhabitants planning restrictions that promote construction in flood risk regions, and, yet at all, inadequate expenditure by sewerage and water firms to modernize systems of Victorian-era, the hazard is only going to get much worse.

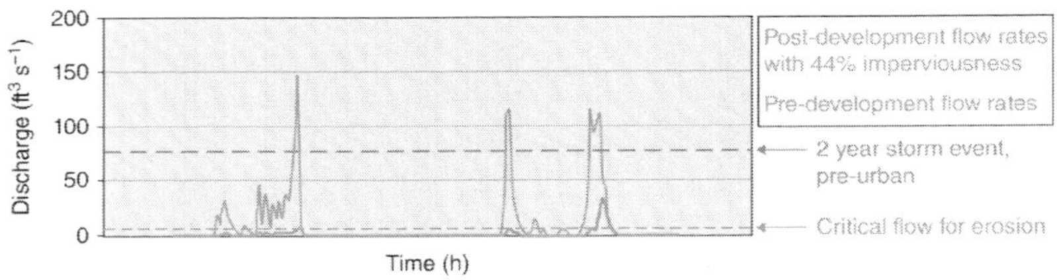

Figure 8.6. California, USA, Thompson Creek Santa Clara Valley, post-, and pre-construction flow conditions, modeled for a 714-acre construction.

Source: www.coastal.ca.gov/nps/watercyclefacts.pdf.

The problem of urban flooding is aggravated by the loss of natural space as a result of increased urbanization; the compaction of pathways and the loosening of planning regulations for several home additions means that water would have nowhere to run. New multi-occupational residential projects are placing further strain on already overcrowded and outdated drainage and sewer systems. The effect of construction on the regional hydrologic cycle is depicted in Figure 8.7.

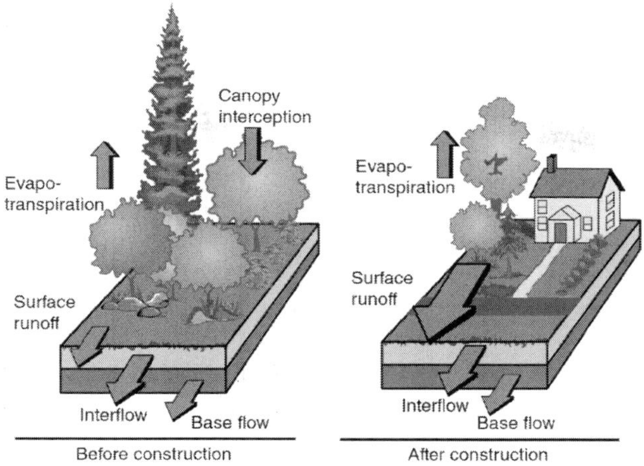

Figure 8.7. Buildings have an impact on the regional hydrologic cycle, as shown by the rise in surface runoff.

Source: https://slideplayer.com/slide/3243685/.

Cole et al. in 2013 describe preliminary findings from a research carried out as part of the British Natural Hazards Partnership, that aims to solve some of the shortcomings of current surface water flooding (SWF) detection systems. The latest NHP prototype SWF strategy develops on CEH's grid to grid model's dynamic gridded surface runoff measurements, which are already used for forecasting fluvial flood operation across the UK. Even though runoff generation in the framework is molded by spatial information sets on the properties of landscape-land cover (e.g., terrain, suburban/urban, geology, and soil, as well as spatially and dynamically varying antecedent moisture of soil, computed via ongoing water accounting inside G2G, this could be a major step forward over current techniques relying on the depth of rainfall. The CEH group worked together with the Health and Safety Laboratory (HSL) to create dynamic maps depicting the effects of water flooding of the surface. The dynamic danger zone is combined with time-changing national effect sets of data on transportation, inhabitants, property, and infrastructure such as the National Human Database from HSL and EA flood danger maps.

8.6. DEVELOPMENT OF FLOODPLAIN

A floodplain is a land area next to a river or stream that runs from the channel's banks to the enclosing valley walls base and floods during periods of heavy discharge. The sideways of meander eroding as it flows downstream creates floodplains. When a river overflows its floods and banks, and layers of alluvium are left behind (silt). The floodplain's floor is progressively built up from these.

Furthermore, local planning officials continue to permit house developments in flood-prone locations. In Wales and England, regional councils permitted at a minimum of 87 planned schemes comprising 560 homes to go forward in 2013 in places at such severe threat of flooding that now the Environment Agency officially objected.

Many cities have historically been constructed on floodplains, because they are very vulnerable to flooding, for a variety of reasons, including the following:

- Development of flat land is simple;
- The availability of clean water;
- The agricultural potential of floodplain land;
- Low-cost transit via railroads and waterways that frequently linked rivers.

Many prosperous countries have lately enacted floodplain building laws. Annual flooding occurrences, on the other hand, are a natural component of the rural economy and regional ecology in a few tropical floodplain locations, such as Mali's Inner Niger Delta, enabling crop production via recessional agriculture. However, in Bangladesh, which is located in the Ganges Delta, the alluvial soil of floodplains benefits is greatly outweighed by regular floods caused by cyclones and yearly monsoon rains, which inflict human death and serious economic damage in this heavily populated region.

For the visualization and investigation of flood inundation zones, computer models offer two-dimensional and one-dimensional simulation methodologies. The Federal Emergency Management Agency (FEMA) in the United States has established guidelines (FEMA, 1995) for engineers, community officials, surveyors, developers, and property owners who may be required to assess the base flood elevations (BFEs) in particular flood potential danger regions assigned as estimated zone A on the maps of FEMA's flood insurance rate published as a component of the Program of National Flood Insurance. Several of the main purposes of these publications is to give a low-cost method of determining BFEs. Likewise, the Environment Agency in the United Kingdom distributes maps of flood risk for all locations on its website.

8.7. ACID RAIN

8.7.1. Basics

Nitrogen oxide and sulfur dioxide emissions combine with water molecules in the atmosphere to form acids, resulting in acid rain (Likens et al., 1972). Dry (gases and acidifying particles and) and Wet (snow, rain, fog, sleet, dew, and cloud water) acidic components are deposited in acid rain. Although acid rain was first discovered in Manchester in 1852, public worry over its effects on vegetation did not spread until the late 1960s (Gorham, 1984 As illustrated in Figure 8.8, pollution generated by human activities is subjected to multiple chemical and physical processes until being deposited again on the Earth's surface.

After carbon dioxide is eliminated, distilled water has a neutral acidity. Acidic liquids have a pH lower than 7, while alkaline liquids have a pH of more than 7. As carbon dioxide and water within in the air react to generate carbonic acid, a kind of weak acid unpolluted rain has an acidic pH (typically no lower than 5.7) according to the reaction given below:

$$H_2O(1) + CO_2(g) \rightleftharpoons H_2CO_3(aq)$$

Carbonic acid then can ionize within the water, creating carbonate and hydronium ions in low concentrations:

$$H_2O(1) + H_2CO_3(aq) \rightleftharpoons HCO_3^-(aq) + H_3O^+(aq)$$

Unpolluted rain, on the other hand, may contain various substances that alter its pH. Nitric acid, which is created by electric discharge within the atmosphere is a common example such as forest fires (Likens et al., 1987) or lightning (Likens et al., 1987).

In the troposphere, nitric oxide (NO) is mostly converted to nitrogen dioxide via interaction with ozone. Deposition of acid as a problem in the environment encompasses acids in addition to H_2CO_3 (Likens and Bormann, 1974).

Wet deposition accounting for a significant portion of overall deposition throughout western Norway, and lower than 15% of total deposition over central Europe (Beilke, 1983). The deposition reported over Europe is mostly due to sources inside Europe (Levy and Maxim, 1987; Whelpdale et al., 1988); relatively little from North America reaches Europe. Metcalfe et al. (1995) found that European land-based sources accounted for 34% of sulfur deposition in 1990, while UK oil refineries and power plants accounted for

46.7%, with natural sources and shipping such as algal blooms accounting for only 4.6%. Wet sulfur deposition in China is 7–130 times greater than in distant southern hemisphere locations (Galloway et al., 1987).

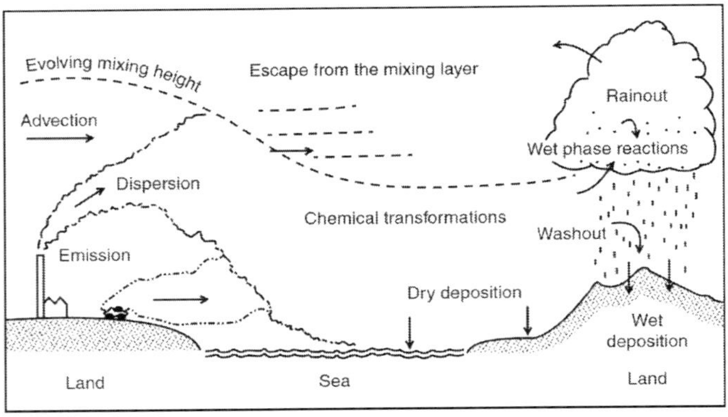

Figure 8.8. Processes that contribute to the deposition of contaminants in the atmosphere, such as acid rain.

Source: https://onlinelibrary.wiley.com/doi/book/10.1002/9781118414965.

Following the Chernobyl disaster in the Soviet Union, radioactive nuclides wet deposition was discovered in numerous European countries, with the quantities of deposited cesium-137 being demonstrated to be directly connected to rainfall intercepting with plume (see for example Clark and Smith, 1988).

8.7.2. Wet Deposition Modelling

The ADMS5 model includes a washout coefficient for modeling wet deposition (Apsley and CERC, 2012). This indicates that all chemical and physical processes utilized in removal of pollution from a plume are expressed by straightforward proportionality among the concentrations of airborne material and removal rates. The assumption include (i) removal procedures are self-contained and (ii) removal procedures are irreversible.

According to the 1st assumption, the strength of the plume can be described as a function of depletion factors and source strength for each operation of removal. When there are species in the release with considerably varied deposition parameters, each must be calculated separately. Therefore, variations in deposition parameters due to the creation of novel isotopes as a result of radioactive decay cannot be predicted but can be modeled and the values of washout coefficient and dry deposition velocity should represent the inventory of the initial isotope. The first assumption rules out the possibility of gases dissolving within raindrops and then emerging from solution near to the ground. This outgassing is not described in any way.

In terms of a washout coefficient, the wet deposition is modeled, and the washout is considered to happen is proportional to the localized concentration all across the plume depth. Except that where large concentrations of CO_2 or SO_2 restrict washout coefficients, they are believed to be homogeneous all across the domain. The amount of pollutant integrated into rainfall when modeled wet deposition using a washout coefficient is ΛC per unit volume per unit time, while C is the concentration of local airborne. Concentrations in raindrops and air, Rainfall rate, aerosol size distributions, raindrop are all factors that influence the washout coefficient. It can be explicitly inputted by a human or predicted by a computer model based on other inputs like rainfall rate.

By integrating via air vertical column and assuming irreversible absorption, the overall rate of wet deposition per unit horizontal area per unit time, F_{wet}, is determined:

$$F_{wet} = \int \wedge C dz \qquad (1)$$

As a result, the plume's strength decreases as it travels further:

$$\frac{dQ}{dx}\bigg|_{wet} = -\int F_{wet} dy \qquad (2)$$

Assumptions are made as follows:

- No distinction is established among below cloud and include scavenging because all plume material is found in or even below rain clouds;
- Pollutant uptake is irreversible, and scavenging of precipitation does not result in material redistribution in the plume;
- Pollutant uptake is irreversible, and scavenging of precipitation does not result in material redistribution in the plume;
- The rate of rainfall is consistent throughout the domain.

The quantity of pollution in the plume that is still present exponentially decays:

$$q_w = e^{-\wedge t_r(x)} \qquad (3)$$

where; t_r is the rain travel time.

The washout coefficient can be set to a constant value, for which case the computed wet deposition is unaffected by the rate of precipitation. In the precipitation absence, the wet deposition will be projected. The possible values for the coefficient of washout are $1.0 \times 10^{-4}\ s^{-1}$. The coefficient of washout can also be modeled just simply as a function of rainfall rate and pollutant species, as shown below:

$$\Lambda = aJ^{b} \qquad (4)$$

where; J is the rate of rainfall ($mm\ h^{-1}$); and a and b are pollutant-specific parameters with a = $1\ 0. \times 10^{-4}$ and b = 0.64 as a default value, respectively (Jylha, 1991). In practice, the restrictive influence of the raindrop pH restricts the uptake of acidic contaminants in large quantities.

8.8. THE URBAN ENVIRONMENT AND SEVERE HYDROMETEOROLOGICAL OCCURRENCES

According to the UN, the world's population will reach 8.3 billion by 2030. Approximately half of the world's population today lives in cities. By 2030, this figure is expected to rise to 61%, resulting in 5.1 billion urban dwellers (United Nations, 2006). In its fourth assessment study published in 2007, the Intergovernmental Panel on Climate Change (IPCC) concluded that, as a result of global warming, a rise in the frequency of severe precipitation events (and hence floods) is "very likely" to occur throughout the 21st century (IPCC, 2007a). In terms of societal vulnerability, the IPCC considers locations where significant urbanization is happening to be among the highly vulnerable (IPCC, 2007b). Extreme weather studies, particularly in urban settings, are crucial for enhancing our knowledge of these occurrences and minimizing their impact in the unpredictable years ahead. The ability to characterize the temporal and spatial features of rainfall during these severe storms has significant implications for analyzing the changing flood threats in metropolitan areas.

Both hydrometeorological and hydrologic factors contribute to the urban environment's vulnerability to flooding, particularly flash flooding. Greater impervious elaboration and cover of the drainage system via the network of storm drain result in high flood peaks and shortened reaction times of urban watershed from a hydrologic standpoint (Leopold, 1968; Graf, 1977; Smith et al., 2002). Hot-season thunderstorms have become the primary cause of flash flooding in several urban areas, which is a significant effect (Ntelekos et al., 2007; Doswell et al., 1996; Baeck and Smith, 1998; Chang,

1976; Smith et al., 2005). As shown below, the urban area could influence the distribution and development of rainfall of hot season thunderstorms systems from a hydrometeorological standpoint.

Modification of rainfall by metropolitan surroundings has been a hot topic of research for the past 40 years, due to the EXperiment of METROpolitan Meteorological (METROMEX) in St. Louis, Missouri (Changnon et al., 1971; Semonin and Changnon, 1974), which sparked a lot of activity. The METROMEX experiment revealed three key findings: (1) an increase in precipitation in the city downwind part in a "fan-shaped" area about 30 kilometers from the center of the city; (2) greater rainfall discrepancies throughout the hot season; and (3) quite noticeable modification in rainfall for storms of high-intensity. The following three mechanisms have been proposed to explain these and similar findings: (1) nonurban and urban areas temperature difference with the urban area to be hotter and therefore establishing convergence centers downwind and over of it (Urban Heat Island, UHI); (2) the excessive roughness of the urban area owing to the infrastructure of an urban area; and (3) the modified chemical characteristics of the urban area.

8.9. WATER POLLUTION AND URBAN AIR

The wet deposition was described in the earlier section. As previously said, this mechanism is crucial for the effects of air pollution within rural and urban areas, including a variety of pollutants like biological molecules, particles, and other dangerous materials. According to the Program of United Nations Environment, over 1 billion people have been vulnerable to outside air pollution each year. Annually, up to 1 million people die prematurely as a result of urban air pollution. In addition, it is expected to cost around 2% of GDP in wealthier countries and 5% in poorer countries. According to a World Health Organization research from 2014, air pollution killed roughly 7 million individuals globally in 2012 (WHO, 2014). Radioactive pollutants, nitrogen oxides, volatile organic compounds, carbon monoxide, persistent free radicals, particulates, hazardous metals, ammonia, chlorofluorocarbons, and Sulfur dioxide are all key air pollutants created by human activities. Raw sewage is a significant health danger since it carries waterborne bacteria which might cause serious sickness in humans. When the related nutrient loading and biological oxygen demand decrease oxygen within the water to amounts very low to maintain life, raw wastewater degrades aquatic ecosystems, posing a hazard to human livelihoods. In 2008, WHO projected that 2 billion individuals worldwide lacked accessibility to better sanitation, having the poorest coverage of 37% in Sub-Saharan Africa, 45% IN Eastern Asia, and 38% in Southern Asia (WHO/UNICEF, 2008). Sewage can be dumped into waterways intentionally using open defecation or pipelines, or inadvertently during heavy rains. When people bathe, wash, or drink in these waterways, they are susceptible to pathogens related to it. Most of these illnesses can live for a prolonged time in aquatic environments. Humans become ill after drinking dirty water, getting it on or in their skin, ears or eyes, or cooking with contaminated water. As a consequence of ingesting dirty water droplets, humans could become ill.

REFERENCES

1. Adams, B. J., & Papa, F., (2000). *Urban Stormwater Management Planning with Analytical Probabilistic Models* (Vol. 1, pp. 1–20). Wiley, Chichester.
2. Apsley, D. D., & CERC, (2012). *Modelling Wet Deposition* (Vol. 1, pp. 1–20). CERC, Cambridge. www.cerc.co.uk/ environmental-software/ assets/data/doc_techspec/CERC_ADMS5_P17_12.pdf.
3. Arnfield, A. J., (2003). Two decades of urban climate research: A review of turbulence exchanges of energy and water, and the urban heat island. *Int. J. Climatol., 23*, 1–26.
4. Baeck, M. L., & Smith, J. A., (1998). Rainfall estimation by the WSR-88D for heavy rainfall events. *Weather Forecasting, 13*(2), 416–436.
5. Beilke, S., (1983). Acid deposition: The present situation in Europe. In: *Acid Deposition* (pp. 3–30). Report EUR8307. Commission of the European Communities.
6. Burian, S. J., & Shepherd, J. M., (2005). Effect of urbanization on the diurnal rainfall pattern in Houston. *Hydrol. Proc., 19*, 1089–1103.
7. Carraca, M. G. D., & Collier, C. G., (2007). Modelling the impact of high-rise buildings in urban areas on precipitation initiation. *Meteorol. Appl., 14*, 149–161.
8. Chandler, T., (1976). The climate of towns. In: Chandler, T. J., & Gregory, S., (eds.), *The Climate of the British Isles* (Vol. 1, pp. 307–329). Longman, London.
9. Chang, C. B., (1976). A case study of excessive rainfall forecasting, *Meteorol. Atmos. Phys., 66*(3–4), 215–227.
10. Changnon, S. A., Huff, F. A., & Semonin, R. G., (1971). METROMEX-an investigation of inadvertent weather modification. *Bull. Am. Meteorol. Soc., 52*(10), 958–967.
11. Changnon, S., & Huff, F. A., (1986). The urban-related nocturnal rainfall anomaly at St. Louis. *J. Clim. Appl. Meteorol., 25*, 1985–1995.
12. Changnon, S., (1981). *METROMEX: A Review and Summary* (Vol. 18, pp. 10–29). Monograph 40, American Meteorological Society, Boston.
13. Changnon, S., Semonin, R. G., & Huff, F. A., (1976). A hypothesis for urban rainfall anomalies. *J. Appl. Meteorol., 15*, 544–560.
14. Choe, J. S., Bang, K. W., & Lee, J. H., (2002). Characterization of surface runoff in urban areas. *Water Sci. Technol., 45*, 249–254.

15. Clark, M. J., & Smith, F. B., (1988). Wet and dry deposition of Chernobyl releases. *Nature*, *322*, 245–249.
16. Cole, S. J., Moore, R. J., Aldridge, T., Lane, A., & Laeger, S., (2013). Real-time hazard impact modelling of surface water flooding: Some UK developments. *Proceedings of the International Conference on Flood Resilience: Experiences in Asia and Europe (ICFR 2013)* (Vol. 1, pp. 1–20). University of Exeter.
17. Collier, C. G., (2006). The impact of urban areas on weather. *Q. J. R. Meteorol. Soc.*, *132*, 1–25.
18. Collier, C. G., Davies, F., Bozier, K. E., Holt, A., Middleton, D. R., Pearson, G. N., Siemen, S., Willetts, D. V., Upton, G. J. G., & Young, R. I., (2005). Dual Doppler lidar measurements for improving dispersion models. *Bull. Am. Meteorol. Soc.*, *86*, 825–838.
19. Cotton, W., & Pielke, R. A., (2007). *Human Impacts on Weather and Climate* (2nd edn., Vol. 2, pp. 1–50). Chapter 5. Cambridge University Press, Cambridge.
20. Deletic, A., Dotto, C. B. S., McCarthy, D. T., Kleidorfer, M., Freni, G., Mannina, G., Uhi, M., Henrichs, M., Fletcher, T. D., Rauch, W., Bertrand-Krajewski, J. L., & Tait, S., (2012). Assessing uncertainties in urban drainage models. *Phys. Chem. Earth*, *42–44*, 3–10.
21. Doswell, C. A. III, Brooks, H. E., & Maddox, R. A., (1996). Flash flood forecasting: An ingredients-based methodology, *Weather Forecasting, 11*, 560–581.
22. FEMA (1995). *Managing Floodplain Development in Approximate Zone A Areas: A Guide for Obtaining and Developing Base (100-Year) Flood Elevations* (Vol. 265, pp. 1–40). FEMA 265. Federal Emergency Management Agency.
23. Galloway, J. N., Dianwu, Z., Jiling, X., & Likenb, G. E., (1987). Acid rain: China, United States, and a remote area. *Science*, *236*, 1559–1562.
24. Goldreich, Y., (1987). Advertent/inadvertent changes in the spatial distribution of rainfall in the central coastal plain of Israel. *Climate Change*, *11*, 361–373.
25. Gorham, E., (1984). Acid rain: An overview. In: Bhumralkar, C. M., (ed.), *Meteorological Aspects of Acid Rain* (Vol. 1, pp. 1–20). Butterworth, London.
26. Graf, W. L., (1977). Network characteristics in suburbanizing streams, *Water Resource. Res., 13*(2), 459–463.

27. Harmon, I. N., Best, M. J., & Belcher, S. E., (2004). Radiative exchange in an urban street canyon. *Boundary-Layer Meteorol., 110*, 301–316.

28. Huff, F. A., & Changnon, S. A., (1972). Climatological assessment of urban effects on precipitation at St. Louis. *J. Appl. Meteorol., 11*, 823–842.

29. IPCC, (2007b), Summary for policymakers, in climate change 2007: Impacts, adaptation and vulnerability. In: Parry, M., et al., (eds.), *Contribution of Working Group II to the Fourth Assessment Report of the Intergovernmental Panel on Climate Change* (Vol. 1, pp. 1–22). Cambridge Uni. Press, U. K., & New York.

30. Jylha, K., (1991). Empirical scavenging coefficients of radioactive substances released from Chernobyl. *Atmos. Environ., 25A*, 263–270.

31. Kishtawal, C. M., Niyogi, D., Tewari, M., Pielke, R. A. Sr and Shepherd, J. M., (2009). Urbanization signature in the observed heavy rainfall climatology over India. *Int. J. Climatol., 30*, 1908–1916.

32. Klysik, K., & Fortuniak, K., (1999). Temporal and spatial characteristics of the urban heat island of Lodz, Poland. *Atmos. Environ., 33*, 3885–3895.

33. Labadie, J. W., Grigg, N. S., & Bradford, B. M., (1975). Automatic control of large-scale combined sewer systems. *J. Environ. Eng. Div. ASCE, 101*, 27–39.

34. Leopold, L. B., (1968). Hydrology for urban planning: A guidebook on the hydrologic of urban land use. *Professional Paper 669, U. S. Geological Survey* (Vol. 1, pp. 1–23).

35. Levy, H. II., & Maxim, W. J., (1987). Fate of US and Canadian combustion nitrogen emissions. *Nature, 328*, 414–416.

36. Likens, G. E., & Bormann, F. H., (1974). Acid rain: a serious regional environmental problem. *Science, 184*, 1176–1179.

37. Likens, G. E., Bormann, F. H., & Noye, M., (1972). Acid rain. In: *Environment Science and Policy for Sustainable Development* (Vol. 14, pp. 33–40). Published online 8 July 2010, Taylor & Francis. doi: 10.1080/00139157.1972.

38. Likens, G. E., Keene, W. C., Miller, J. M., & Galloway, J. N., (1987). Chemistry of precipitation from a remote terrestrial site in Australia. *J. Geophys. Res., 92*, 13299–13314.

39. Lin, C.-Y., Chen, W.-C., Chang, P.-L., & Sheng, Y.-F., (2011). Impact of the urban heat island effect on precipitation over a complex geographic

environment in Northern Taiwan. *J. Appl. Meteorol. Climatol., 50*, 339–353.

40. Metcalfe, S. E., Whyatt, J. D., & Derwent, R. G., (1995). A comparison of model and observed network estimates of sulfur deposition across Great Britain for 1990 and its likely source attribution. *Q. J. R. Meteorol. Soc., 121*, 1387–1411.
41. Ntelekos, A. A., Smith, J. A., & Krajewski, W. F., (2007). Climatological analyses of thunderstorms and flash floods in the Baltimore metropolitan region. *J. Hydrometeorol., 8*(1), 88–101.
42. Oke, T. A., (1982). The energetic basis of the urban heat island. *Q. J. R. Meteorol. Soc., 108*, 1–24.
43. Oke, T. A., (1987). *Boundary Layer Climates* (2nd edn., Vol. 2, pp. 1–30). Routledge, London.
44. PCC, (2007a), Summary for policy makers, in climate change 2007: The physical science basis. In: Solomon, S., et al., (eds.), *Contribution of Working Group I to the Fourth Assessment Report of the Intergovernmental Panel on Climate Change* (Vol. 1, pp. 1–13). Cambridge Uni. Press, U. K., & New York.
45. Schellart, A., Ochoa, S., Simoes, N., Wang, L.-P., Rico-Ramirez, M., Lignon, S., Duncan, A., Chen, A. S., Keedwell, E., Djordjevic, S., Savic, D. A., Saul, A., & Maksimovic, C., (2011). *Urban Pluvial Flood Modelling with Real Time Rainfall Information: UK Case Studies* (Vol. 12, pp. 1–40). 12th International Conference on Urban Drainage, Porto Alegre, Brazil.
46. Schmitt, T. G., Thomas, M., & Ettrich, N., (2004). Analysis and modelling of flooding in urban drainage systems. *J. Hydrol., 299*, 300–311.
47. Schueler, T., (1987). *Controlling Urban Runoff: A Practical Manual for Planning and Designing Urban Best Management Practices* (Vol. 1, pp. 1–34). MWCOG, Washington, DC.
48. Semonin, R. G., & Changnon, S. A., (1974). METROMEX-summary of 1971–1972 results. *Bull. Am. Meteorol. Soc., 55*(2), 95–100.
49. Shepherd, J. M., & Burian, S. J., (2003). Detection of urban-induced rainfall anomalies in a major coastal city. *Earth Interactions, 7*, 1–17.
50. Smith, F. B., (1984). The fate of airborne pollution. Section 1. In: *Acid Rain, Report 14* (pp. 1–13). Watt Committee on Energy.
51. Smith, J. A., Baeck, M. L., Meierdiercks, K. L., Nelson, P. A., Miller, A.

J., & Holland, E. J., (2005). Field studies of the storm event hydrologic response in an urbanizing watershed. *Water Resour. Res., 41*, W10413, doi: 10.1029/2004WR003712.

52. Smith, J. A., Baeck, M. L., Morrison, J. E., Sturdevant-Rees, P., Turner-Gillespie, D. F., & Bates, P. D., (2002). The regional hydrology of extreme floods in an urbanizing drainage. *J. Hydrometeorol., 3*(3), 267–282.
53. UN, (2001). *World Population Monitoring 2001* (pp. 1–29). Department of Economic and Social Affairs, Population Division, UN E.01.xii.17. United Nations, New York.
54. United Nations, (2006). *World Population Prospects: The 2006 Revision-Analytical Report* (Vol. III, pp. 1–22). Technical Report No. 26.
55. Whelpdale, D. M., Eliassen, A., Galloway, J. N., Dovland, H., & Miller, J. M., (1988). The transatlantic transport of sulfur. *Tellus, 408*, 1–15.
56. WHO, (2014). *Seven Million Premature Deaths Annually Linked to Air Pollution* (Vol. 1, pp. 1). World Health Organization. https://www.who.int/news/item/25-03-2014-7-million-premature-deaths-annually-linked-to-air-pollution.
57. WHO/UNICEF, (2008). *Progress on Drinking Water and Sanitation: Special Focus on Sanitation* (Vol. 1, pp. 1–23). International Year of Sanitation, WHO, Geneva, UNICEF, New York. https://apps.who.int/iris/handle/10665/43931.

INDEX

S

T

V

W